7th International Symposium on Gas Transfer at Water Surfaces 2015

AA005627

IOP Conference Series: Earth and Environmental Science
Volume 35

Seattle, Washington, USA
18 – 21 May 2015

Editors:

Andrew T. Jessup
William E. Asher

ISBN: 978-1-5108-2447-8
ISSN: 1755-1307

Printed from e-media with permission by:

Curran Associates, Inc.
57 Morehouse Lane
Red Hook, NY 12571

Some format issues inherent in the e-media version may also appear in this print version.

Copyright© (2015) by the Institute of Physics
All rights reserved. The material featured in this book is subject to IOP copyright protection, unless otherwise indicated.

Printed by Curran Associates, Inc. (2016)

For permission requests, please contact the Institute of Physics at the address below.

Institute of Physics
Dirac House, Temple Back
Bristol BS1 6BE UK

Phone: 44 1 17 929 7481
Fax: 44 1 17 920 0979

techtracking@iop.org

Additional copies of this publication are available from:

Curran Associates, Inc.
57 Morehouse Lane
Red Hook, NY 12571 USA
Phone: 845-758-0400
Fax: 845-758-2633
Email: curran@proceedings.com
Web: www.proceedings.com

Table of contents

Volume 35

7th International Symposium on Gas Transfer at Water Surfaces
18–21 May 2015, Seattle, Washington, USA

Accepted papers received: 18 April 2016
Published online: 20 May 2016

Preface

Papers

IOP Conf. Series: Earth and Environmental Science **35** (2016) 011001 doi:10.1088/1755-1315/35/1/011001

PREFACE

The 7th International Symposium on Gas Transfer at Water Surfaces was held 18–21 May 2015 in Seattle, Washington. These symposia, which have occurred approximately every five years since 1983, bring together the international scientific and engineering community investigating the mechanisms, parameterizations, and applications of gas transfer at water surfaces. The focus in part is on the physical processes that govern the gas flux, which include air entrainment (by breaking waves, flow over hydraulic structures, or direct injection), turbulence (generated by mechanical agitation, wind, wave breaking, rainfall, or currents), and the effect of natural and synthetic surfactants. Of equal importance to the treatment of the forcing mechanisms are the discussions on how to model or parameterize air-water gas transfer relevant to biogeochemical cycling on global, regional, and local scales. Furthermore, because the mechanisms controlling the gas flux also govern the air-sea flux of heat and momentum, the conference is attended by many studying a wide range of air- and water-side mixing processes that occur at or near the ocean surface. This provides opportunities for cross-fertilization of ideas between the classical air-sea interaction community and those focused on biogeochemical cycling.

The first symposium was held at Cornell University, Ithaca, New York in 1983 [1]. The second symposium took place in Minneapolis in 1990 [2] and the third symposium was held at the University of Heidelberg in Heidelberg, Germany in 1995 [3]. The fourth symposium took place in Miami in 2000 [4] and the fifth symposium was held in Liège, Belgium in 2005 in conjunction with the 37th International Liège Colloquium on Ocean Dynamics [5]. The sixth symposium was held in Kyoto, Japan in 2010 [6].

Gas transfer at air-water interfaces encompasses a wide variety of research, including fundamental fluid dynamics, biogeochemistry, and oceanography across a range of spatial and temporal scales. For example, the exchange of greenhouse gases between the atmosphere and natural bodies of water is critical to characterize and quantify global climate change. The establishment of a regular series of international symposia with a periodicity of five years was motivated by the increasing societal interest in the consequences of gas transfer at air-water interfaces, the size and productivity of the research community, and the wide geographic distribution of active researchers. The symposium has established a reputation as an important and influential venue for presenting and disseminating research progress to the community. The five-year period between gatherings ensures that significant progress since the last occurrence will be reported.

The research problems involve extremely challenging and complex issues associated with turbulent flow over a wide range of spatial and temporal scales that include the effects of wind and currents, breaking waves, rainfall, and two-phase flow effects due to the presence of bubbles generated by breaking waves and raindrops. The complexity is increased by the occurrence of both natural and anthropogenic surface films, which can modify the chemical and physical characteristics of the interface. The challenge of understanding gas transfer and the effect of gas exchange on biogeochemistry has led to the need for multidisciplinary, collaborative efforts and the development of a wide variety of innovative observational and experimental techniques. As many dedicated past attendees can attest, the symposia have fostered productive and long-lasting collaborations that have made new and pivotal contributions over the past three decades.

Content from this work may be used under the terms of the Creative Commons Attribution 3.0 licence. Any further distribution of this work must maintain attribution to the author(s) and the title of the work, journal citation and DOI.
Published under licence by IOP Publishing Ltd

IOP Conf. Series: Earth and Environmental Science **35** (2016) 011001 doi:10.1088/1755-1315/35/1/011001

The objectives of the meeting were:

- to provide a balanced view of state-of-the-art research to understand, quantify, and parameterize gas transfer at water surfaces by disseminating and evaluating progress since the 2010 meeting
- to determine how best to use these new results to better understand and parameterize biogeochemical cycling over global, regional, and local scales
- to foster multidisciplinary interaction and collaboration within the gas transfer community

The topics covered by previous symposia encompass field, laboratory, and modeling investigations of the physics governing air-water transfer, development and testing of parameterizations for use in global climate models, remote sensing techniques, and assessment of the global distribution of air-sea flux of CO_2 and other radiatively important trace gases. In addition to these core topics, the 2015 symposium included the topics of ocean acidification due to increased CO_2 uptake, the impact of reduced sea ice extent on gas transfer in high-latitude regions, and studies in coastal regions, rivers, and lakes. The Seattle meeting was attended by 108 participants from North America, Asia, and Europe, with over 100 oral and poster presentations. The papers in this volume provide a sampling of the wide variety of field, laboratory, and modeling research results presented at the meeting.

We gratefully acknowledge the contributions of the local organizing and scientific committees in coordinating the meeting, providing guidance, and assisting in the review process for the papers in this volume.

Local Organizing Committee: A. Jessup (Chair), W. Asher, (co-Chair), E. D'Asaro, A. Horner-Devine, C. McNeil, S. Mecking, and A. Schweiger, University of Washington, USA.

Scientific Committee: A. Jessup (Chair) and W. Asher (Co-Chair), University of Washington, USA; J. Boutin, CNRS/LOCEAN-IPSL, France; M. Donelan, University of Miami, USA; C. Garbe, University of Heidelberg, Germany; J. Gulliver, University of Minnesota, USA; H. Herlina, Karlsruhe Institute of Technology, Germany; D. Ho, University of Hawaii, USA; S. Komori, Kyoto University, Japan; L. Miller, Institute of Ocean Sciences, Canada; W. McGillis, Columbia University, USA; P. Nightingale, Plymouth Marine Laboratory, England; A. Rutgersson, Uppsala University, Sweden; E. Saltzman, National Science Foundation, USA; J. Shutler, University of Exeter, England; R. Stanley, Woods Hole Oceanographic Institution, USA; R. Wanninkhof, NOAA Atlantic Oceanographic and Meteorological Laboratory, USA; B. Ward, National University of Ireland, Ireland; C. Zappa, Columbia University, USA.

The symposium would not have been possible without the generous contributions of our sponsors. We gratefully acknowledge support from the Applied Physics Laboratory and the Department of Civil and Environmental Engineering at the University of Washington, the National Aeronautics and Space Administration, the National Science Foundation, Ocean Carbon and Biogeochemistry, and US CLIVAR.

We are grateful for the assistance in organizing the meeting from Syd Fredrickson of the University of Washington Conference Management and in preparing this volume from Lisa Day-Mercer and Brian Rasmussen from the Applied Physics Laboratory, University of Washington.

This volume is dedicated to the memory of Sergei A. Kitaigorodskii and Edgar L. Andreas. Both of these dedicated scientists participated in several of the symposia and made unique and lasting contributions to the field of gas transfer, as noted in the tributes that follow.

We look forward to the 8th International Symposium on Gas Transfer at Water Surfaces, which will be hosted by Drs. Jamie Shutler of the University of Exeter and Phil Nightingale of the Plymouth Marine Laboratory.

Andrew T. Jessup
William E. Asher

References

[1] Brutsaert W and Jirka G H 1984 *Gas Transfer at Water Surfaces* (Hingham, MA: Reidel) 639 pp

[2] Wilhelms S C and Gulliver J S 1991 *Proceedings of the Second International Symposium on Gas Transfer at Water Surfaces* (New York: ASCE) 797 pp

[3] Jähne B and Monahan E C 1995 *Air-Water Gas Transfer, Selected Papers from the Third International Symposium on Air-Water Gas Transfer* (Hanau, Germany: Aeon Verlag & Studio) 900 pp

[4] Donelan M A *et al* 2002 *Gas Transfer at Water Surfaces* (Washington, DC: AGU) 383 pp

[5] Borges A V and Wanninkhof R 2007 37th International Liege Colloquium on Ocean Dynamics, Liege, Belgium, May 2–6, 2005, 5th International Symposium on Gas Transfer at Water Surfaces *J. Mar. Syst.* **66** 1-308

[6] Komori S *et al* 2011 *Gas Transfer at Water Surfaces 2010* (Kyoto: Kyoto University Press) 594 pp

In Memoriam: Sergei A. Kitaigorodskii, 1934–2014

K Kahma

Finnish Meteorological Institute, FI-00561 Helsinki, Finland

Professor Sergei A. Kitaigorodskii, one of the pioneers of air-sea interaction studies, passed away on 4 December 2014 in Helsinki, at the age of 80. He had a clear vision of the fundamental problems of marine boundary layer physics. His bold ideas opened new paths more than a half century ago, and they continue to stimulate active research. In particular, he made significant progress on the difficult problem of estimating gas transfer rates through gas–liquid interfaces and contributed to many of the Gas Transfer at Water Surfaces conferences.

Kitaigorodskii was born in Moscow on 13 September 1934. His father, Professor Alexander Kitaigorodskii, was a well-known physicist. Sergei Kitaigorodskii completed his undergraduate studies at Moscow State University in 1956, and he continued his postgraduate studies on the theory of turbulent mixing at the Academy of Sciences, Moscow, where he was awarded his Ph.D. from the Institute of Physics of the Atmosphere in 1960.

As a post-doctoral fellow at the Institute of Oceanology of the Academy of Sciences, he formulated the Kitaigorodskii similarity hypothesis for the wind-generated wave spectrum in 1961. He later recounted how, as a young student, he had been impressed by the simple and at the same time fundamental results of O. M. Phillips, and he had wanted to follow that path by combining it with the Russian tradition, the Kolmogorov similarity hypothesis of turbulence. The Kitaigorodskii hypothesis immediately showed its value when W. J. Pierson Jr. and L. Moskowitz were able in 1964 to collapse their empirical data into a single dimensionless curve, and to formulate the Pierson–Moskowitz spectrum for fully developed seas. At the same time Kitaigorodskii continued his studies of the atmospheric side and showed how swell could influence the marine boundary layer, a topic upon which his ideas remain important.

In 1962 Kitaigorodskii became a Senior Scientist at the institute and turned his main interest back to turbulence of the oceanic mixed layer. Here he again made fundamental contributions. These, combined with his earlier studies of waves and the air-sea boundary layer, were summarized in his thesis for the Russian degree of Doctor of Science in 1968. This work was the basis of a monograph, *Physics of Air-Sea Interaction*, published in 1970 and translated into English in 1973; it was used widely as a textbook. In the same year, 1973, he received the Rosenstiel gold medal, and in 1978 the Liege University Award.

Between 1968 and 1977 Kitaigorodskii was head of the Laboratory of the Physics of Ocean-Atmosphere Interaction in the Institute of Oceanology.

During the 1970s Kitaigorodskii established contacts with the Finnish geophysics community. His Finnish wife had completed her term as a press correspondent in Moscow, and she returned to Helsinki with their twin daughters. The Soviet system allowed Kitaigorodskii to visit his family every other year, and he used this opportunity to lecture and conduct research at several Finnish institutes.

Content from this work may be used under the terms of the Creative Commons Attribution 3.0 licence. Any further distribution of this work must maintain attribution to the author(s) and the title of the work, journal citation and DOI.
Published under licence by IOP Publishing Ltd

Sylvain Joffre describes the lectures in Helsinki:

> *Sergei's trademark was his ability to boil down issues and problems encountered during the research process to the core physical processes and questions. Sergei's lectures were also always a source of inspiration. They were like a Hollywood movie, alternating drama and comedy, and climaxing suddenly with a simple, beautiful, and enlightening physical explanation. The resulting blackboard, full of equations and graphs spreading in all directions and into every corner, was in itself an artistic collage. Sergei was not a practical person in everyday life, and though he was aware of it, it did not bother him. Nor did it have a negative impact on the way others looked at him; on the contrary, this perhaps contributed to creating his legend, and we all looked up to him as a genuinely great scientist, with an image that perfectly suited that role.*

My own introduction to air-sea interaction was mainly through these lectures. I therefore took it as self-evident that Sergei's ideas described things as they were, even though not all had been proven. The extent of his visionary ideas was revealed to me later in my studies. I found that there were alternative views to his, and then decades later empirical evidence showed that, in many cases, things were indeed very close to what Sergei had envisioned.

In 1978 Kitaigorodskii moved abroad. He was first a visiting professor at the Max Planck Institute for Meteorology, the Johns Hopkins University, and Cornell University; then a lecturer at the University of Copenhagen in 1979 – 1985. In 1980 – 1992 he was Professor of Oceanography at the Johns Hopkins University. During this period he collaborated with John Lumley at Cornell University, Kristina Katsaros at the University of Washington, and with the research group led by Mark Donelan at the Canada Centre for Inland Waters. Donelan's comprehensively instrumented field experiments in Lake Ontario, which included new instruments specifically developed for this campaign, demonstrated the important role of breaking waves for subsurface turbulence — an issue that Sergei had emphasized years before in his Helsinki lectures.

In 1992 Sergei returned to Europe. He was first a visiting professor at the Finnish Meteorological Institute and the Finnish Institute of Marine Research; from 1994 to 2001 he was at the Institute of Oceanology of the Russian Academy of Sciences. He spent his last years living in Helsinki near his family.

Sergei was active to the very end. He completed the book that he had been working on for decades, *Five Discoveries by Harald Sverdrup. An Introduction to Physical Oceanography*. He gave seminars and wrote papers — mostly summaries of his previous work. But when chess championship matches were happening, everything else was postponed. Sergei loved chess and he was an excellent player.

IOP Conf. Series: Earth and Environmental Science **35** (2016) 011003 doi:10.1088/1755-1315/35/1/011003

In Memoriam: Edgar L. Andreas, 1946–2015

E C Monahan and P Vlahos

Department of Marine Sciences, University of Connecticut, Groton, CT 06340, USA

On a day early last autumn, Ed Andreas went out, as was his wont, for a run from his home in Lebanon, New Hampshire. While on this run, Ed, a contributor to this volume and a collaborator and friend of many of us in the air-sea exchange community, suffered a cardiac event, from which he succumbed on 30 September 2015.

Mention of Ed Andreas will immediately bring to many readers' minds his protracted effort to incrementally improve the existing parameterizations of the sea surface spray flux (e.g., Andreas *et al.* 1995; Andreas 1998; Andreas 2001; Andreas *et al.* 2015) and his recent efforts to combine this modeling with concepts from atmospheric chemistry to assess the potential significance of spray-mediated air-sea gas flux (Andreas *et al.*, this volume). Since 2007 Ed worked on these topics from his home, which served as his outpost of NorthWest Research Associates. But it should be noted that many of his earlier papers dealt with the Arctic marine boundary layer, and sea-air fluxes at high latitudes (e.g., Andreas *et al.* 1979; Makshtas *et al.* 1986), reflecting his almost 30-year long tenure (1978–2006) at the U.S. Army's Cold Regions Research and Engineering Laboratory of the U.S. Army Corps of Engineers, located just down the road from Lebanon, in Hanover, New Hampshire.

Having been born in Sterling, Illinois, Ed went off to nearby Knox College for his B.S. in Physics, and then on to Michigan State University for an M.S. in Physics, before deserting the Midwest for Corvallis, where he earned his Ph.D. in Physical Oceanography at Oregon State University in 1977. While Ed certainly became an active member of our global scientific community, he never lost his Midwestern virtues. He could write well, and received numerous awards for his technical communication, but he was often refreshingly candid in speech. Ed would on occasion provide honest "unvarnished" critiques of his and his collaborators' efforts, but he did not hesitate to praise others' efforts in print. Ed was a principled person, and this was manifest in his professional dealings with others. Working on a manuscript with Ed was always, as it should be, an ultimately enlightening intellectual exercise.

His approach to any task combined both the rigor of the physicist with the practicality of an engineer, and as a result of his efficient nature, he has left our air-sea interaction community with a trove of no less than 140 valuable scientific publications, with others yet to appear.

Ed Andreas was a Fellow of the American Meteorological Society and of the Royal Meteorological Society (UK). He was recipient of the Antarctica Service Medal, and of numerous performance awards from the Cold Regions Research and Engineering Laboratory.

This brief tribute just touches upon Ed Andreas as a scientist. But Ed was a truly multi-faceted individual, as is apparent to any reader of the posthumous biographical sketch that appeared in his local

Content from this work may be used under the terms of the Creative Commons Attribution 3.0 licence. Any further distribution of this work must maintain attribution to the author(s) and the title of the work, journal citation and DOI.
Published under licence by IOP Publishing Ltd

newspaper (Jurgens 2015), which you may want to access. We would like to acknowledge with gratitude the biographical materials we received from Ed's immediate family, and from his former associates at NorthWest Research Associates, Inc. in Redmond, Washington.

References

Andreas E L 1998 A new sea spray generation function for wind speeds up to 32 m/s *J. Phys. Oceanogr.* **28** 2175-2184

Andreas E L and Emanuel K A 2001 Effects of sea spray on tropical cyclone intensity *J. Atoms. Sci.* **58** 3741-3751

Andreas E L *et al* 1995 The spray contribution to net evaporation from the sea: A review of recent progress *Bound-Lay. Meteorol.* **72** 3-52

Andreas E L *et al* 2015 An improved bulk air-sea surface flux algorithm, including spray-mediated transfer *Q. J. Roy. Metorol. Soc.* **141** 642-654

Andreas E L *et al* 1979 The turbulent heat flux from Artic leads *Bound-Lay. Meteorol.* **17** 57-91

Jurgens R 2015 A life, 1946–2015: A brilliant scientist and an explorer of the physical, spiritual and psychological world *Valley News* (archived at www.vnews.com/Archives/2015/11/alifeedgarandreas-rj-vn-112315)

Makshtas A P *et al* 1986 Small-scale air-sea interaction in the vicinity of Maud Rise *Information Bulletin of the Soviet Antarctic Expeditions* **108** 67-71 (Petrograd: Arctic and Antarctic Research Institute) *in Russian*

IOP Conf. Series: Earth and Environmental Science **35** (2016) 011004 doi:10.1088/1755-1315/35/1/011004

Peer review statement

All papers published in this volume of *IOP Conference Series: Earth and Environmental Science* have been peer reviewed through processes administered by the proceedings Editors. Reviews were conducted by expert referees to the professional and scientific standards expected of a proceedings journal published by IOP Publishing.

Content from this work may be used under the terms of the Creative Commons Attribution 3.0 licence. Any further distribution of this work must maintain attribution to the author(s) and the title of the work, journal citation and DOI.
Published under licence by IOP Publishing Ltd

IOP Conf. Series: Earth and Environmental Science **35** (2016) 012001 doi:10.1088/1755-1315/35/1/012001

A mixing length model for the aqueous boundary layer including the effect of wave breaking on enhancing gas transfer

M A Donelan[1] and A V Soloviev[1, 2]

[1]Rosenstiel School of Marine and Atmospheric Science, University of Miami, Miami, Florida, USA
[2]Oceanographic Center, Nova Southeastern University, Dania Beach, Florida, USA

E-mail: mdonelan@rsmas.miami.edu

Abstract. A mixing length model for air-water gas transfer is developed to include the effects of wave breaking. The model requires both the shear velocity induced by the wind and the integrated wave dissipation. Both of these can be calculated for tanks and oceans by a full spectrum wave model. The gas transfer model is calibrated, with laboratory tank measurements of carbon dioxide flux, and transported to oceanic conditions to yield air-sea transfer velocity versus wind speed.

1. Introduction

Mixing length theory has been applied to the air boundary layer to model the transport of water soluble compounds such as water vapour, aerosols, SO_2, SO_3 and NH_3 [1, 2]. Wind-generated waves on the air-sea interface enter this problem only in that they modify the wind stress ($\tau = \rho_a u_{*a}^2$) and the principal scaling velocity for the mass flux, i.e., the friction velocity u_{*a}. The wind profile very near the surface is controlled by molecular viscosity, ν, and is linear in height, z ; whereas away from the surface (wall units, $zu_*/\nu = z^+ > 100$), the profile is proportional to the logarithm of height. Mixing length parameterization of the eddy viscosity/diffusivity provides the simplest way to model the profiles (of wind/contaminant) with the observed smooth transition from linear to logarithmic in flow over a smooth wall [3]. To deal with transitional and rough boundaries, [4] and [1] proposed additive mixing lengths, introduced by the surface roughness, for momentum and mass respectively; thereby providing a general model for air-phase limited gas transfer through smooth, transitional and rough surfaces [5].

Sparingly soluble gases such as O_2, N_2, CO_2, CO, CH_4 and the inert gases are under water-phase control. The aqueous boundary layer is also ventilated by the water turbulence generated by the wind stress acting on the surface but, unlike the air-phase, it is also mixed directly by the turbulence produced in wave breaking. Here the turbulent velocities injected may be an order of magnitude larger than those due to the wind stress. This leads to ventilation of the aqueous boundary layer and entrainment of air bubbles to depths comparable to the wave height. Both of these are related to wave breaking dissipation, D_w, which is the integral over all waves of the dissipation source function in wave models. The bubble-mediated transfer further depends on the solubility of the entrained gas, being greater for lower solubility [6]. While it is generally accepted that wave breaking enhances the transfer of gases that are limited in the water-phase, it is not known to what extent or how D_w affects the mass transfer velocity.

Content from this work may be used under the terms of the Creative Commons Attribution 3.0 licence. Any further distribution of this work must maintain attribution to the author(s) and the title of the work, journal citation and DOI.
Published under licence by IOP Publishing Ltd

IOP Conf. Series: Earth and Environmental Science **35** (2016) 012001 doi:10.1088/1755-1315/35/1/012001

The principal problem in the experimental determination of the effect of wave breaking on gas transfer lies in the extreme difficulty of measuring D_w and u_{*W}. However, a wave model with properly formulated source functions for wave dissipation and wind input can be used to estimate both D_w and u_{*W}. Such a wave model has been developed and proven for gravity waves [7] and is extended to include capillary-gravity and capillary waves for the full spectrum calculations required here. With this wave model, careful and precise laboratory measurements of mass transfer velocities can be transported to the field.

2. The mixing length model

Modeling the average properties of the boundary layers at the air-water interface requires closure of the Reynolds equations. The simplest closure represents the diffusivity of the turbulence in analogy to molecular viscosity:

$$(\nu+\nu_t)\frac{dU}{dz}=u_*^2$$

where constant flux layers on both sides of the interface are assumed. In describing the turbulent boundary layer, ν_t, the eddy viscosity is taken to be $\nu_t = \kappa u_* z$, leading to the usual logarithmic "law of the wall"; κ is von Kármán's constant. In the context of gas transfer this is not adequate because the main resistance to gas transfer comes from the thin diffusive layers on either side of the interface. The concept of mixing length is particularly useful as it provides a physical picture of the eddy sizes that are dominant in the mixing process. It may be thought of as the average vertical distance eddies travel before losing their characteristics (velocity, temperature, etc.) to the average surrounding flow. Formally, it is related to the eddy viscosity:

$$\nu_t = l_u^2\frac{dU}{dz}$$

where in the logarithmic layer: $l_u = \kappa z$.

Clearly, there is a need to have a smooth transition between the viscous sub-layer, having a linear profile, and the logarithmic layer. This was suggested by [8] and modified by [4], and used to good effect in several descriptions of boundary layers on smooth solid walls:

$$l_u = \kappa z[1-\exp(-zu_*/13\nu)]^2 .$$

The corresponding mixing length for passive scalars:

$$l_g = Sc_t^{-\frac{1}{2}}\kappa z[1-\exp(-zu_* Sc^{0.33}/13\nu)]^2$$

where Sc_t is the turbulent Schmidt number and is a function of the Monin–Obukhov stability index, ς. Thus the universal behaviour of diabatic boundary layers is built into the mixing length. The molecular Schmidt number dependence, $Sc^{0.33}$, is determined from the smooth flow values (at U_{10} = 2 m/s) of mass transfer velocity of water vapour and carbon dioxide.

To deal with flows that are rough or transitional – those in which the thickness of the viscous boundary layer is less than the height of the roughness elements – [2] introduced a non-zero mixing length at the surface, l_{us}:

IOP Conf. Series: Earth and Environmental Science **35** (2016) 012001 doi:10.1088/1755-1315/35/1/012001

$$l_u = l_{us} + \kappa z[1 - \exp(-zu_*/13\nu)]^2$$

which leads to the following (fitted) relationship between the roughness length, z_0 and l_{us}:

$$l_{us} = z_0[3.35\exp(-0.1\{0.25/\mathrm{Re}_*\}^4)\exp(-0.96\{\mathrm{Re}_*/0.25\}^{0.33}) + 0.449]; \quad \mathrm{Re}_* > 0.11$$

$$l_{us} = 0; \quad \mathrm{Re}_* \leq 0.11$$

where $\mathrm{Re}_* = z_0u_*/v$ is the roughness Reynolds number. l_{us} increases from zero to $1.61z_0$ at $\mathrm{Re}_* = 0.25$; then decreases exponentially to $0.449z_0$ in fully rough flow.

The ratio of surface mixing lengths for gases to that for momentum is necessarily a function of the other relevant non-dimensional numbers, i.e., the molecular Schmidt number and the roughness Reynolds number:

$$l_{gs} = 0.684Sc^{-0.5}\,\mathrm{Re}_*^{-0.47}\,l_{us}.$$

where the empirical coefficients are set by comparison with the data of [9] and [15].

3. Wave breaking and enhancement of gas transfer of water-side limited gases

Wave breaking is a source of vertical turbulent energy to the aqueous boundary layer that acts to disrupt the thin viscous sub-layer near the surface. The integrated (over all wavenumbers) variance dissipation rate, D_w, provides an unambiguous connection with the effect of wave breaking on the mixing length. The full spectrum wave model is used to obtain D_w in the laboratory tank, the ocean, and the large circular flume of Heidelberg University's Aeolotron (figure 1).

D_w is a strong function of wind speed and a weak function of fetch – compare 230 km (black circles) with 4 km (black squares) oceanic fetches. The dissipation rates for the laboratory tank (length, width, height = 32.2 m, 0.76 m, 0.85 m) with water depth of 0.225 m, as used in the measurements of CO_2 flux, are given for a fetch of 14.3 m (black diamonds). The tank dissipation rates are much smaller (up to a factor of 30) than those computed for oceanic fetches. They permit transporting the precise laboratory measurements to the field. In order to compare with data from Heidelberg University's Aeolotron [10], D_w was computed for its circular shape (green diamonds). Note that the Aeolotron's (water depth of 1 m) dissipation rates are much larger than the linear tank's at low wind speeds because of the homogenous quasi-infinite fetch of the Aeolotron. At higher wind speeds the waves become depth-limited in both cases; the deeper Aeolotron being about three times larger.

D_w (units of viscosity, m^2/s) can be thought of as an eddy viscosity associated with wave breaking. Flow mixing effects of wave breaking, such as mass transfer of gases and emulsification of oils, can be parameterized in terms of a "wave dissipation number," $W_d = D_w/v$.

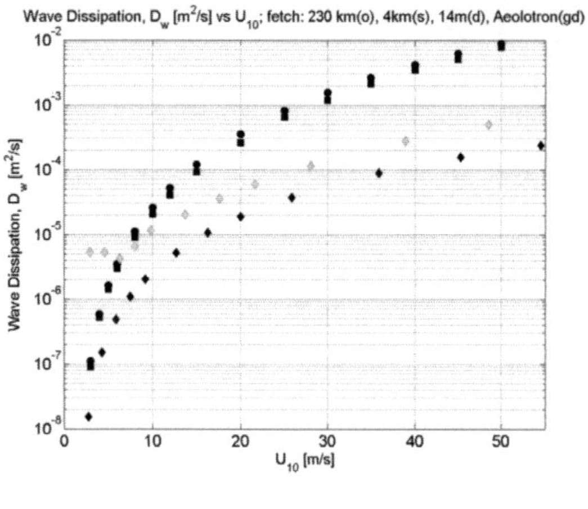

Figure 1. Modelled wave dissipation versus wind speed. Ocean at 230 km (black circles); ocean at 4 km (black squares); linear tank (black diamonds); circular flume (green diamonds).

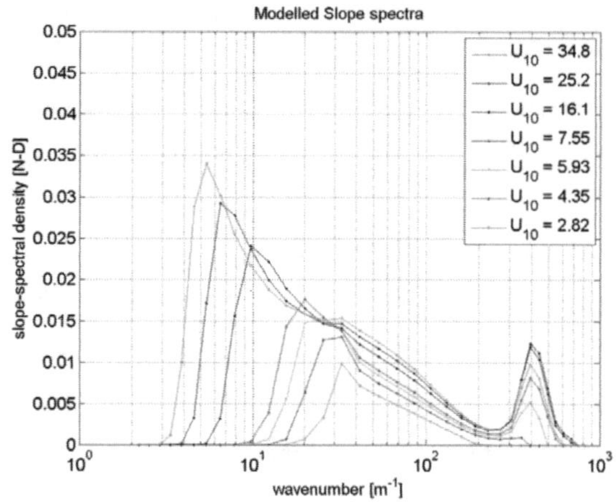

Figure 2. Modelled slope spectrum in the linear tank for various wind speeds in m/s.

Figure 2 shows the modelled slope spectra versus wavenumber for various U_{10} in m/s. The spectra have two modes: a wide band extending across the gravity wave spectrum to wavelengths of about 10 cm, and a narrow band centred on the capillary-gravity minimum phase speed waves (wavelength 1.73 cm in clean fresh water). These small breaking waves (micro-breakers) mix the thin aqueous boundary layer very effectively and provide a scaling length for the mixing length due to breaking.

The additive mixing lengths due to wave dissipation, l_{uw} and l_{gw}, may now be determined by comparison of the mixing length model with the laboratory CO_2 transfer velocities [11]. Figure 3 compares the measured and modelled transfer velocity versus tank centreline wind speed, with and without D_w.

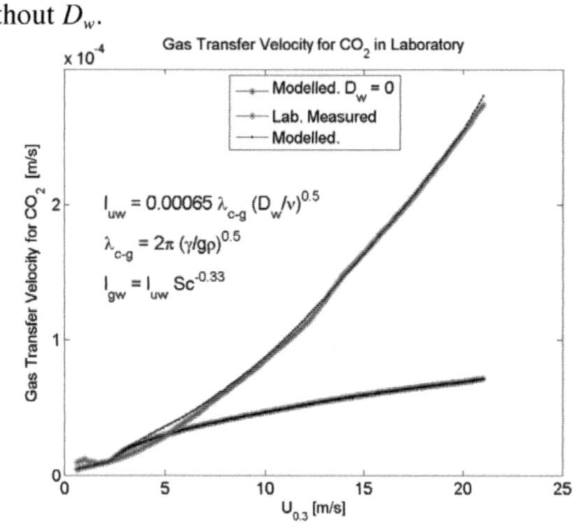

Figure 3. Measured and modelled gas transfer velocity in linear tank versus centerline wind speed.

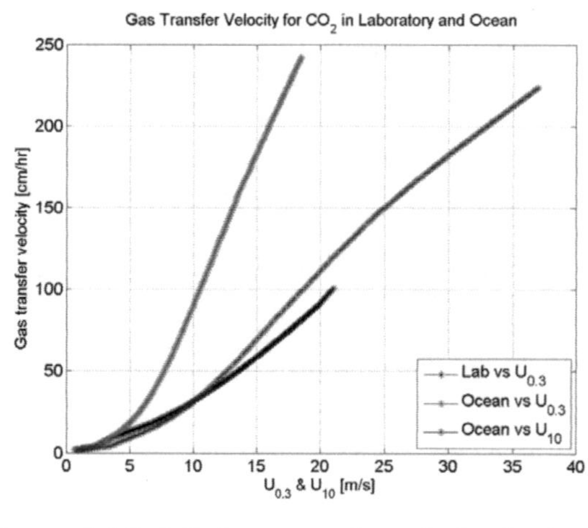

Figure 4. Modelled gas transfer velocity in the laboratory and ocean versus wind speed.

l_{uw} is scaled with the center capillary-gravity wavelength, $\lambda_{c-g} = 2\pi(\gamma/g\rho)^{0.5}$, where γ is the surface tension and ρ the water density. To obtain a close fit between measured (red line) and modelled (blue line) gas transfer velocities, the additive wave dissipation mixing lengths for velocity, l_{uw}, and gas, l_{gw}, are:

$$l_{uw} = 0.00065\lambda_{c-g}(D_w / \nu)^{0.5}$$
$$l_{gw} = l_{uw}Sc^{-0.33}$$

The coefficient and exponents have been determined empirically. The black line, representing the modelled gas transfer velocity without any wave dissipation, illustrates the enormous impact of wave dissipation on the transfer of sparingly soluble gases.

Figure 4 compares the modelled gas transfer velocity in the laboratory (black line) and ocean (red line) using the corresponding wave model calculated D_w and u_*. The effect of the increased wave dissipation in the ocean is about a factor of two when plotted against $U_{0.3}$. While the laboratory transfer velocity shows a quadratic dependence on $U_{0.3}$, the oceanic transfer velocity is quadratic at low U_{10}, and approximately linear above 20 m/s.

4. Schmidt number dependence

The Schmidt number dependence of the mixing length model is tested by comparison with measurements in the same laboratory tank (figure 5) at Schmidt numbers of 600 [11] and higher [12]. Measurements are in red; model in blue.

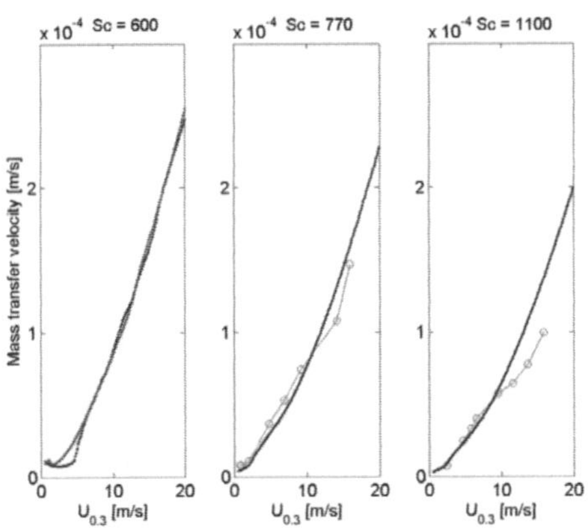

Figure 5. Measured and modelled gas transfer velocity in linear tank versus centerline wind speed for various Schmidt numbers.

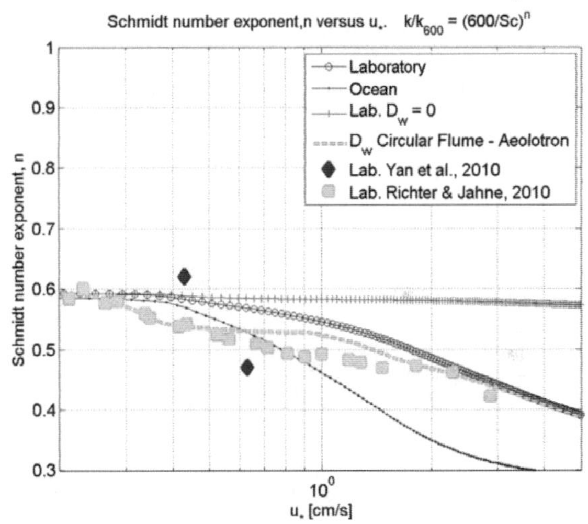

Figure 6. Measured and modelled Schmidt number exponent versus friction velocity in the water.

The Schmidt number exponent, n, is defined by the gas transfer velocity = aSc^{-n}. This exponent is plotted against friction velocity in figure 6 for the model with D_w values of zero (red line), linear tank (black line), Aeolotron (green dashed line) and ocean (blue line). The Aeolotron measured values (green squares [10]) are in close agreement with the modelled (green) line in the low and high thirds of log(u_*). Overall, both measurements and model show a downward trend with increasing u_* and D_w, showing decreased sensitivity to Schmidt number as the surface turbulence intensity increases. Applying laboratory measured Schmidt number dependencies to the field will lead to gross errors especially at high winds.

5. Carbon dioxide transfer on the oceans

The CO_2 transfer laboratory experiments [11] also explored the effect of ambient current on the gas transfer velocity. They found that the imposed current contributed 1.03% to the effective friction velocity. This yields a simple prescription for calculating the gas transfer velocity in the presence of ambient (at depth of 10 m) currents. Figure 7 shows the modelled CO_2 transfer velocity versus U_{10} for various ambient currents plotted with the measurements of [13] and the model of [14]. The mixing length model can be used to calculate the oceanic transfer velocities for any sparingly soluble gas provided that the wave dissipation rate, the friction velocity and the ambient current are known.

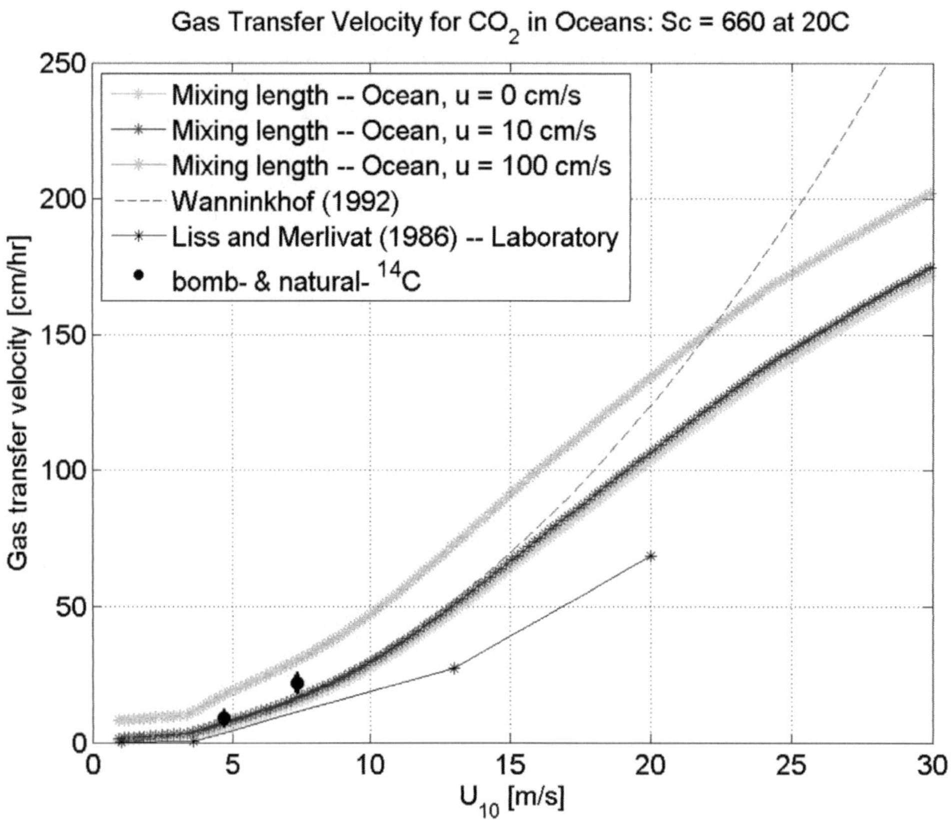

Figure 7. Modelled CO_2 transfer velocity in oceans at 20°C versus wind speed, with "u" the ambient current.

6. Summary

The mixing length model provides a means of calculating mass transfer velocities for 'contaminants' that are limited in either phase. The various empirical coefficients that define the relationship between mixing lengths for momentum and passive scalars were set by comparing with the wide Schmidt number range data of [15] for Schmidt number dependence and to the evaporation data of [9] for the roughness Reynolds number dependence. The model, thus established for the air boundary layer, is applied to the water boundary layer with the important role of wave breaking in mixing the sub-layer included through the integrated wave dissipation rate. Ocean transfer velocities of carbon dioxide are found to be less than linear in the 10-m neutral stability speed at high winds, not quadratic or cubic as has been suggested. They are further enhanced by the upper ocean ambient currents. The model is general and can be used for all sparingly soluble gases and any gas-liquid interface.

References

[1] Donelan M A 1983 *A mixing length model for heat and mass transfer at the air-water interface*(Burlington, Ontario, Canada: National Water Research Institute)

IOP Conf. Series: Earth and Environmental Science **35** (2016) 012001 doi:10.1088/1755-1315/35/1/012001

[2] Kitaigorodskii S A and Donelan M A 1984 Wind-wave effects on gas transfer *Gas Transfer at Water Surfaces*, ed W Brutsaert and G H Jirka (Dordrecht: D. Reidel) pp 147-170
[3] Reichardt H 1940 *Z. Angew. Math. Mech.* **20** 297
[4] Riley D S *et al* 1982 *Bound.-Layer Meteor.* **22** 209
[5] Donelan M A 1990 Air-sea interation *The Sea: Ocean Engineering Science* vol 9 ed B LeMehaute and D M Hanes (New York: John Wiley and Sons) pp 239-292
[6] Woolf D K 1997 Bubbles and their role in gas exchange *The Sea Surface and Global Change*, ed P S Liss and R A Duce (Cambridge: Cambridge University Press) pp 173-205
[7] Donelan M A *et al* 2012 *J. Geophys. Res.* **117** C00J23 doi:10.1029/2011JC007787
[8] Van Driest E R 1956 *J. Aero. Sci.* **23** 1007
[9] Jeong D *et al* 2012 *J. Atmos. Sci.* **69** 2733
[10] Richter K E and Jahne B 2011 A laboratory study of the Schmidt number dependency of air-water gas transfer *Gas Transfer at Water Surfaces* ed S Komori et al. (Kyoto: Kyoto University Press) pp 322-332
[11] Ocampo-Torres F J *et al* 1994 *Tellus* **46 B** 16
[12] Alaee M *et al* 1995 Wind and wave effects on mass transfer velocities of halomethanes and SF_6 measured in a gas transfer flume *Air-Water Gas Transfer*, ed B Jahne and E C Monahan (Hanau: AEON Verlag & Studio) pp 617-626
[13] Liss P S and Merlivat L 1986 Air-sea gas exchange rates: Introduction and synthesis *The Role of Air-Sea Exchange in Geochemical Cycling* ed P Buat-Menard (Dordrecht: Reidel) p 113
[14] Wanninkhof R 1992 *J. Geophys. Res.* **97** 7373
[15] Moller V and Schumann 1970 *J. Geophys. Res.* **75** 3013

Assessing the utility of satellite-based whitecap fraction to estimate sea spray production and CO_2 transfer velocity

To the memory of Dr. Edgar L. Andreas

M D Anguelova
Remote Sensing Division, Naval Research Laboratory, Washington, DC, USA

maggie.anguelova@nrl.navy.mil

Abstract. The utility of a satellite-based whitecap database for estimates of surface sea spray production and bubble-mediated gas transfer on a global scale is presented. Existing formulations of sea spray production and bubble-mediated CO_2 transfer velocity involve whitecap fraction parametrization as a function of wind speed at 10 m reference height $W(U_{10})$ based on photographic measurements of whitecaps. Microwave radiometric measurements of whitecaps from satellites provide whitecap fraction data over the world oceans for all seasons. Parametrizations $W(U_{10})$ based on such radiometric data are thus applicable for a wide range of conditions and can account for influences secondary to the primary forcing factor, the wind speed. Radiometric satellite-based $W(U_{10})$ relationship was used as input to: (i) the Coupled Ocean-Atmosphere Response Experiment Gas transfer (COAREG) algorithm to obtain CO_2 transfer velocity and total CO_2 flux; and (ii) the sea spray source function (SSSF) recommended by Andreas in 2002 to obtain fluxes of sea spray number and mass. The outputs of COAREG and SSSF obtained with satellite-based $W(U_{10})$ are compared with respective outputs obtained with the nominal $W(U_{10})$ relationship based on photographic data. Good comparisons of the gas and sea spray fluxes with direct measurements and previous estimates imply that the satellite-based whitecap database can be useful to obtain surface fluxes of particles and gases in regions and conditions difficult to access and sample in situ. Satellite and in situ estimates of surface sea spray production and bubble-mediated gas transfer thus complement each other: accurate in situ observations can constrain radiometric whitecap fraction and mass flux estimates, while satellite observations can provide global coverage of whitecap fraction and mass flux estimates.

1. Introduction

Direct measurements of near-surface ocean processes under high winds remain problematic due to the difficulty of operating surface vessels and in situ instrumentation under gale conditions. Such measurements are necessary to better constrain parameterizations of near-surface ocean processes and improve predictions of weather and climate models. Whitecaps are the surface expression of various air-sea processes enhanced by wave breaking. Whitecap fraction W—defined as the fraction of the sea surface covered with sea foam—is thus a forcing variable suitable for parameterizing surface fluxes associated with breaking waves and whitecaps such as surface production of sea spray aerosol and bubble-mediated gas exchange across the air-sea interface. Whitecap fraction is usually obtained from photographs by applying an intensity threshold $W(I)$ to discriminate whitecaps from the surrounding water. Photographic $W(I)$ data have been used extensively to derive parameterizations of W as a function of wind speed at 10 m reference height $W(U_{10})$. Surface fluxes of particles and gases can be estimated

Content from this work may be used under the terms of the Creative Commons Attribution 3.0 licence. Any further distribution of this work must maintain attribution to the author(s) and the title of the work, journal citation and DOI.
Published under licence by IOP Publishing Ltd

globally by combining a $W(U_{10})$ expression with global maps of U_{10}. However, available $W(U_{10})$ relationships do not capture the full variability of the oceanic whitecaps because many other factors beside U_{10} affect the formation and behavior of the bubbles comprising the whitecaps.

Experiments have shown strong correlation of whitecap fraction and gas transfer velocity with microwave brightness temperature T_B [1]. Observations of ocean surface T_B carry information about the whitecaps because T_B responds directly to the surface roughness and sea foam as they are affected by U_{10}, wave field history, and seawater properties (e.g., temperature, salinity, and presence of organic compounds) [2]. Attenuation of the electromagnetic radiation at microwave frequencies (e.g., 6–40 GHz) by atmospheric gases, water vapors, and clouds is well characterized and affords all-weather information on oceanic whitecaps from satellite-borne radiometers. Use of satellite-based whitecap fraction to parameterize gas exchange and other surface fluxes of mass and energy can thus alleviate the difficulties of direct in situ measurements of surface fluxes under high winds and potentially improve the flux accuracy by accounting for the whitecap variability caused by secondary factors.

An algorithm obtaining W from satellite observations, developed at the Naval Research Laboratory, combines T_B data from WindSat radiometer, a model for atmospheric attenuation at microwave frequencies, and physically-based models for the emissivity of rough sea surface and emissivity of foam [3]. The whitecap database built with this algorithm compiles $W(T_B)$ values for entire year 2006 matched in time and space with data for the wind vector (speed and direction), wave field (e.g., significant wave height and peak wave period), and environmental parameters (e.g., sea surface temperature and atmospheric stability) from different sources [4]. This database has proved useful in analyzing and quantifying the variability of W [5].

By their measuring principle, microwave radiometers provide the total whitecap fraction $W(T_B)$, including foam generated during active wave breaking and residual foam left behind by these breaking waves [6]. The total whitecap fraction W is important for estimating the sea spray aerosol production and heat exchange between the ocean and the atmosphere. However, the active portion W_A is necessary for evaluating turbulent processes such as mixing of the upper ocean, gas exchange, ocean ambient noise, and spray-mediated intensification of tropical storms [7]. Sensitivity of microwave frequency to foam thickness allows association of W_A with $W(T_B)$ at 10 GHz (hereafter referred to as W_{10}) and W with $W(T_B)$ at 37 GHz (referred to as W_{37}) [5, 6]. Because this is a crude separation of active and residual whitecaps, we have developed a scaling factor $R = W_A/W$, which, when applied to satellite-based data $W(T_B)$, can provide data for W_A on a global scale [8]. The procedure separating active and residual whitecaps uses the Philips concept for breaking crest length distribution [9] and energy dissipation rate ε [10].

Here, we present estimates of carbon dioxide (CO_2) transfer velocity and the sea spray production flux, and assess differences due to use of $W(U_{10})$ parameterizations based on photographic or radiometric W data. Use of W_A values instead of W and the applicability of satellite-based whitecap fraction values for estimating surface fluxes of particles and gases are discussed.

2. Methods

2.1. CO_2 transfer velocity and flux

A physically-based algorithm for bulk meteorological and gas transfer fluxes has been developed using data from the Coupled Ocean-Atmosphere Response Experiment (COARE) [11]. Consistently adding new observations from different field campaigns, the COARE working group has been steadily improving the representation of the meteorological and gas fluxes in the COARE algorithm [11]. Data from three Gas Exchange (GasEx) field campaigns have been used to implement the COARE gas transfer algorithm (designated COAREG) [12].

The COAREG algorithm has taken the representation of the gas transfer velocity k away from simple power laws in terms of U_{10} and currently partitions k into waterside and airside components [12]. Each component is formulated in terms of forcing variables (e.g., U_{10} or friction velocity u_*), gas solubility, and gas diffusivity (Schmidt number). The waterside term k_w includes molecular-turbulent (or direct) transfer and bubble-mediated transfer k_b. The bubble-driven term k_b is taken from Woolf [13] and involves the wind speed parameterization of whitecap fraction $W(U_{10})$ among other variables. Two parameters, A and B, in k_w are used as tuning parameters, their values adjusted every time new gas flux observations (or process parameterizations) are available. Importantly, parameters A and B can be used to judge how well the physical formulations in the COAREG algorithm represent the modeled processes and forcing factors. Specifically, because COAREG represents the gas transfer physics on the basis of the similarity hypothesis, the empirical constants A and B should be the same for all gasses.

The latest version COAREG3.1 [12] has been used in this study with meteorological data from the Southern Ocean (SO) GasEx field campaign [14]. The whitecap fraction parameterization used in the bubble-mediated transfer velocity k_b is that developed by Monahan and O'Muircheartaigh [15] (hereafter referred to as MOM80):

$$W(U_{10}) = 3.84\times10^{-6}U_{10}^{\;3.41} \tag{1}$$

COAREG3.1 uses $A = 1.6$ and $B = 1.8$, values adjusted to fit data from all three GasEx campaigns and wind stress representation with tangential and wave components [12]. Algorithm version COAREG3.1 produces a good fit to field observations using the same A and B values for two gases, CO_2 and DMS (dimethyl sulfide) (e.g., figure 4 in [12]). One aspect to be addressed in future COAREG versions is to reconcile predictions for the total transfer velocity (normalized to a fixed Schmidt number of 660) k_{660} from COAREG3.1 to eddy correlation (atmospheric) gas flux measurements during GasEx campaigns [14] and to the quadratic dependence $k_{660}(U_{10})$ based on dual tracer (oceanographic) gas flux measurements [16].

2.2. Sea spray production flux

Sea spray production flux is necessary for estimating sea spray aerosol loadings in the atmosphere and the spray-mediated heat fluxes. The number or mass of small sea spray aerosols (radii at formation $r_0 \leq$ 1.6 μm) are needed when modeling the direct scattering of shortwave radiation and modifications of cloud properties by sea salt particles. Loadings of large sea spray aerosols ($r_0 > 1.6$ μm) are needed to evaluate the contribution of sea spray to enthalpy flux (the sum of sensible and latent heat fluxes) and its role for intensification of tropical storms. For chemical transport models, surface and volume provided by the sea spray particles are needed to assess heterogeneous chemical reactions occurring on or within sea spray droplets.

Sea spray production flux is quantified with a sea spray source function (SSSF), which predicts the rate at which sea spray droplets at a given size dF/dr_0 are produced at the sea surface. SSSF has two main dependencies: (i) size dependence, which gives the shape of the SSSF over the range of sea spray sizes; and (ii) wind speed dependence, which gives the magnitude of the flux. In this study, we use the SSSF recommended by [17]. In this SSSF, the wind speed dependence is introduced via MOM80 whitecap fraction given with (1). The size dependence is presented with the size distributions of Monahan *et al.* [18] for small ($0.8 \leq r_0 \leq 1.6$ μm) and Andreas [19] for large ($1.6 \leq r_0 \leq 500$ μm) droplets.

2.3. Whitecap fraction

There are whitecap fraction data available now from three different measuring methods (section 1), namely photographic data $W(I)$, radiometric data $W(T_B)$, and data from energy dissipation rate using a $W_A(\varepsilon)$ expression. Figure 1 (solid line) shows the MOM80 parameterization $W(U_{10})$ given with (1). The available photographic data $W(I)$ (circles, [2, their table 2]) cluster around the MOM80 $W(U_{10})$ parameterization. Data for wind speed and wave spectra from buoy 41001 have been used to obtain ε and then calculate active whitecap fraction $W_A(\varepsilon)$ (diamonds in figure 1). Values $W_A(\varepsilon)$ compare reasonably well with photographic $W_A(I)$ data (crosses in figure 1). Satellite-based data $W(T_B)$ at frequencies of 10 and 37 GHz, matched in time and space with data from buoy 41001, are also shown in figure 1 (squares and triangles). Both photographic $W(I)$ and radiometric $W(T_B)$ data are one to two orders of magnitude larger than active whitecap fraction data $W_A(I)$ and $W_A(\varepsilon)$. Overall, whichever method of separating W and W_A is used, most values of active whitecap fraction lay below the MOM80 $W(U_{10})$ curve.

The question how best to use the satellite-based $W(T_B)$ data has been reasoned by Albert *et al.* [20]; i.e., should we use the data directly or should we use them to first derive a new $W(U_{10})$ parameterization. Using the $W(T_B)$ values directly limits the uncertainty to that of estimating $W(T_B)$ from satellite observations and does not add uncertainty from deriving an expression for $W(U_{10})$. Alternatively, a new $W(U_{10})$ parameterization can be used with readily available wind speed data. Importantly, such a parameterization will be globally applicable because the whitecap database from which it is derived covers the full range of meteorological conditions encountered over the world oceans for all seasons.

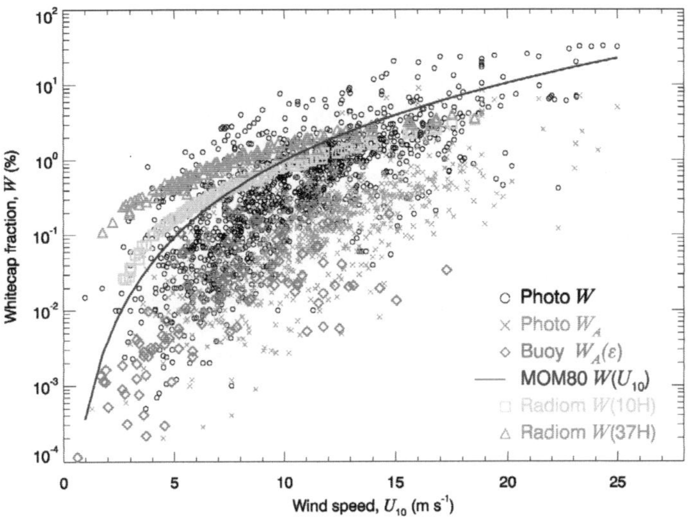

Figure 1. Whitecap fraction obtained with different methods as a function of wind speed at 10 m reference height U_{10}: total whitecap fraction W from photographs (circles); active whitecap fraction from photographs W_A (crosses); total whitecap fraction from satellite observations of brightness temperature at 10 and 37 GHz, horizontal (H) polarization (squares and triangles); active whitecap fraction from energy dissipation rate ε (diamonds); parameterization $W(U_{10})$ based on photographic W data (solid line).

The large number of W values from the whitecap database ensures low error in the derivations of the $W(U_{10})$ expression from satellite observations. In this study, we used the $W(U_{10})$ parameterizations based on $W(T_B)$ values for 10 and 37 GHz [5]:

$$W_{10} = 4.6\times10^{-3}\times U_{10}^{2.26};\quad 2 < U_{10} \le 20 \text{ m s}^{-1}, \tag{2a}$$

$$W_{37} = 3.97\times10^{-2}\times U_{10}^{1.59};\quad 2 < U_{10} \le 20 \text{ m s}^{-1}, \tag{2b}$$

Analysis of the $W(T_B)$ database implied that the influence imparted to secondary factors on whitecap fraction is likely well captured by quadratic wind speed dependence [20]. The wind exponents of 2.26 for W_{10} and 1.56 for W_{37} in (2) thus reflect the effects of additional factors well. Importantly, the deviations of the wind exponents in (2) from the quadratic exponent account for specifics pertinent to active and residual whitecaps, namely more wind and turbulence influence on W_{10} and more bubble property influence on W_{37}.

For this study, we have run COAREG3.1 and SSSF first with (1), the nominal choice for whitecap fraction $W(U_{10})$ in the respective formulations. We then ran COAREG3.1 and SSSF with (2) at all other parameters the same, i.e., changing only the $W(U_{10})$ relationship. Comparing the results from the two runs, we gain insights for the differences in CO_2 transfer velocity, CO_2 flux, and sea spray production due to using a whitecap fraction with different (or absent) representation of secondary influences. The difference for a variable x obtained once with the nominal (photographic) $W(U_{10})$ and then with the new (radiometric) $W(U_{10})$ relationships is quantified as percent change (PC) defined as:

$$PC = \Delta x/x = 100\left(x_{new} - x_{nominal}\right)/x_{nominal} \quad (3).$$

3. Results and discussion

Figure 2 shows bubble-mediated transfer velocity k_b obtained with (1) and (2). For wind speeds ranging from 0.5 to ~ 21 m s^{-1}, k_b values vary from < 10^{-4} to ~ 100 cm h^{-1}. The satellite-based values for k_b are larger than the nominal k_b values for wind speeds below ~ 10 m s^{-1} and smaller for high winds (above 15 m s^{-1}). This is caused by wind speed exponents lower in (2) than that of MOM80 in (1). There is little significance to the disagreement at low wind speeds, with $\Delta k_b/k_b$ >> 200% from (3), considering that the wind speed for the onset of whitecapping ranges from 3 to ~ 6 m s^{-1} [2, their table 1]. The k_b values obtained with (2a) at 10 GHz are closer to the nominal k_b values from COAREG3.1 than k_b values obtained with (2b) at 37 GHz. Because W_{10} from (2a) represents predominantly active whitecaps with only partial contribution from residual whitecaps (section 1), W_{10} is more suitable for gas exchange estimates than W_{37}. For U_{10} > 10 m s^{-1}, the use of (2a) leads to underestimation of k_b compared to the nominal COAREG3.1 k_b by no more than 60%, i.e., from (3) $\Delta k_b/k_b$ = (–60, 0)%.

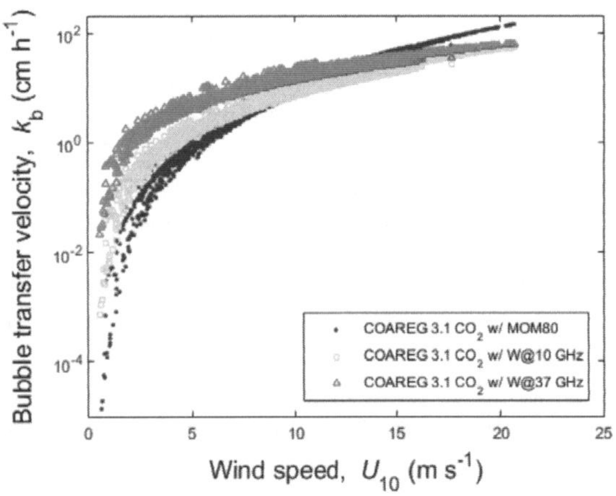

Figure 2. Bubble-mediated transfer velocity k_b as a function of wind speed U_{10}.

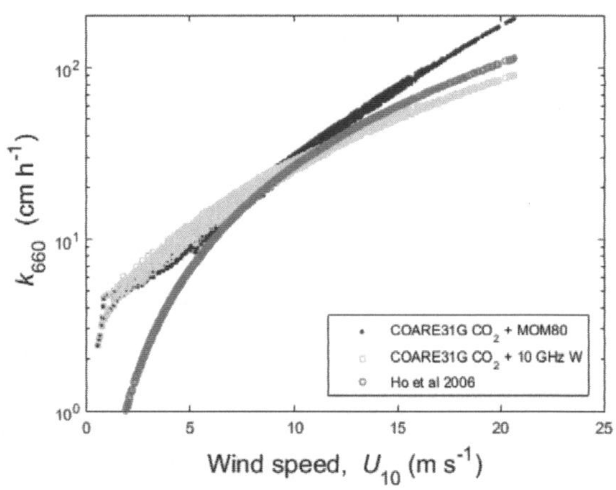

Figure 3. Normalized transfer velocity k_{660} as a function of wind speed U_{10}.

Figure 3 shows how the difference in k_b translates to a difference in the (normalized) total transfer velocity k_{660} (recall section 2.1). The satellite-based k_{660} values compare closely with the nominal COAREG3.1 k_{660} for wind speeds up to about 10 m s^{-1}. At higher winds, satellite-based k_{660} values are closer to the quadratic relationship obtained from dual tracer observations than to the COAREG3.1 values tuned to the eddy correlation observations (section 2.1). Comparison of time series of the total CO_2 flux F_{CO2} (figure 4) shows that the use of (2a) instead of (1) yields differences $\Delta F_{CO2}/F_{CO2}$ from +13% to –53% (figure 5) with an average underestimation of ~ – 6%. The largest difference (around yearday 100) is for a period with the highest wind speed. As expected, the use of (2b) yields overestimation of the total CO_2 flux with $\Delta F_{CO2}/F_{CO2}$ of ~ + 30% on average.

The estimates shown in figures 2–5 are obtained using the total whitecap fraction values $W(I)$ and $W(T_B)$. As the data shown in figure 1 suggest, if we use active whitecap fraction values $W_A(I)$ or $W_A(\varepsilon)$ instead of W, we would obtain much lower bubble-mediated and total normalized CO_2 transfer velocity and flux, by at least a factor of 10. Such low estimates would be in stark contradiction with the COAREG values, which have been adjusted to direct measurements. This implies that, though W_A is considered more suitable for gas exchange occurring during the turbulent phase of the wave breaking, some mix of upper limit values of W_A and lower limit values of W may represent the gas transfer well.

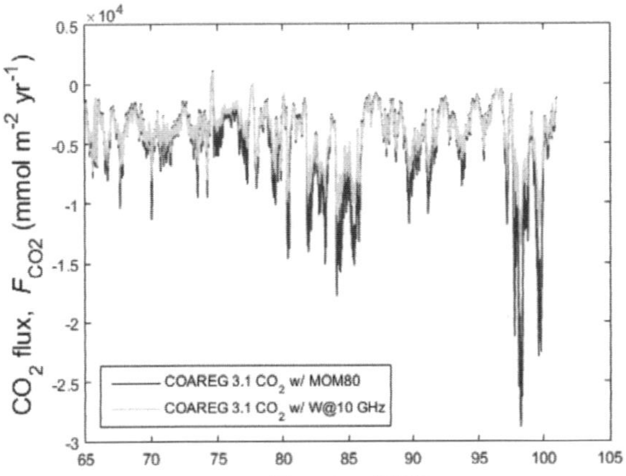

Figure 4. Time series of air-sea CO_2 flux.

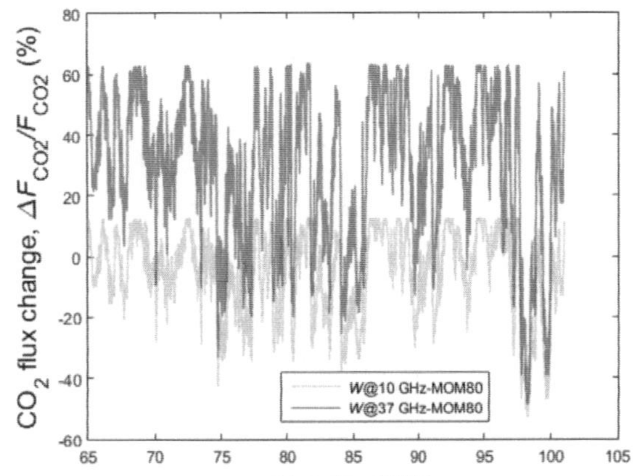

Figure 5. Time series of the percent change of air-sea CO_2 flux due to change of $W(U_{10})$ relationship.

Figure 6 shows the size-resolved sea spray number flux at wind speed 15 m s^{-1}. The vertical (dotted) line at 1.6 μm marks the portions of the size spectrum modeled as small and large particles (section 2.2). Using the same size distributions in SSSF but changing the $W(U_{10})$ relationship from (1) to (2), we obtain lower sea spray fluxes. The sea spray flux obtained with W_{10} (dashed line) is by a factor of 2 lower than the flux estimated with MOM80 parameterization (dash-dot line). This is an upper limit of the photographic versus radiometric differences because the flux dF/dr_0 at 37 GHz obtained with (2b) (solid line) has values between those of dF/dr_0 obtained with (1) and (2a). Considering that both active and residual whitecaps contribute to the bubble-mediated production of sea spray droplets, W_{37} is more suitable for estimates of spray-driven heat fluxes than W_{10}. Figure 7 shows a difference map of sea spray fluxes at r_0 = 10 μm calculated with (1) and (2b) for one day (27 March 2006). According to the map, at this size, sea spray number flux dF/dr_0 obtained with (1) is higher (lower) at high (low) latitudes than dF/dr_0 obtained with (2b) by up to ± 300 particles.

The annual rate of total (size integrated) sea spray mass flux dF for March 2006 is shown in figure 8. From such maps, we estimated global mean of sea spray production of 5.64 Pg yr^{-1}. This estimate is comparable to the values of Jaeglé *et al.* [21] and Sofiev *et al.* [22] (4.86 and 5.87 Pg yr^{-1}, respectively), and it differs by less than a factor of 2 from the values reported by Norris *et al.* [23] and Grythe *et al.* [24] (3.25 and 8.91 Pg yr^{-1}, respectively). While the differences can be partially explained with flux integration over different sizes in different studies, it is noteworthy that these previous estimates are based either on direct observations (e.g., Norris *et al.* [23] uses direct eddy correlation measurements) or on SSSFs modified to better account for additional factors (e.g., SST in Jaeglé *et al.* [21]).

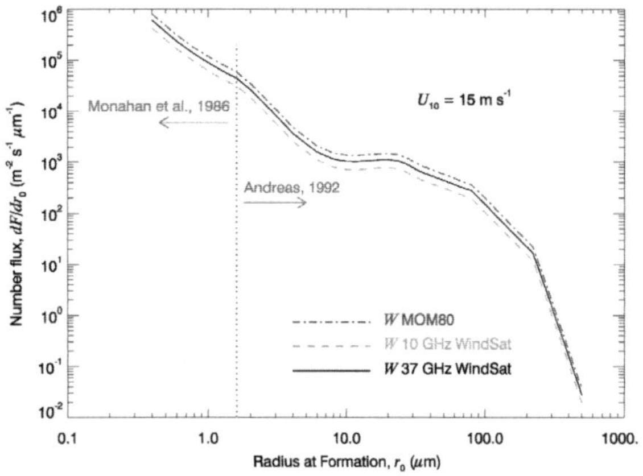

Figure 6. Size-resolved sea spray number flux at wind speed 15 m s^{-1} obtained with different formulations of the whitecap fraction wind speed dependence $W(U_{10})$. The vertical line marks small and large sea spray radii at formation r_0 calculated with different size distributions.

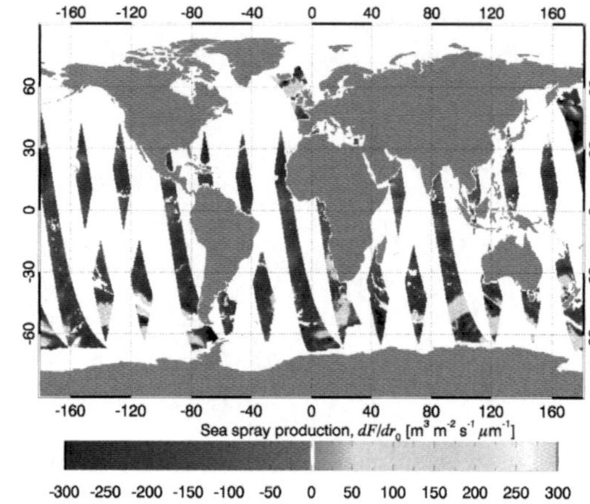

Figure 7. Daily map for 27 March 2006 (grid cells of 0.5°·0.5°) of the difference of sea spray number fluxes at r_0 = 10 μm obtained with $W(U_{10})$ of MOM80 and WindSat: positive anomalies (light colors) show that dF/dr_0(MOM80) > dF/dr_0(WindSat).

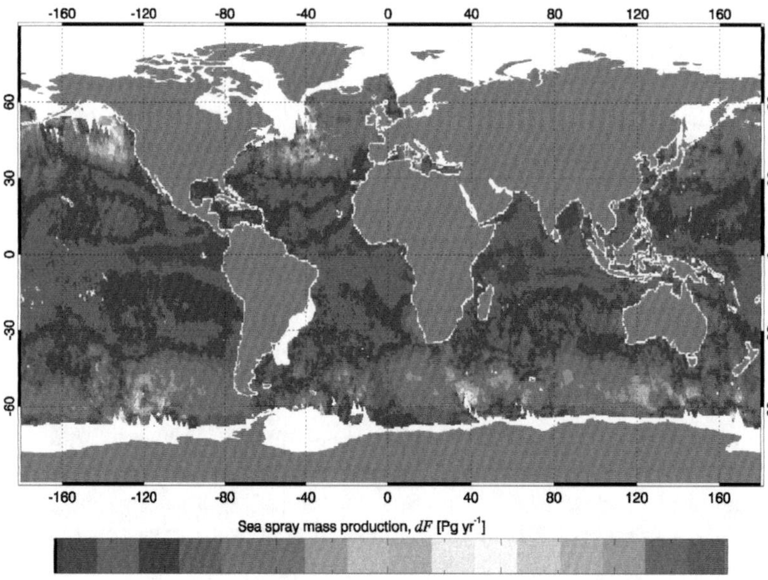

Figure 8. Monthly map (March 2006) of annual rate of total (integrated by size) sea spray mass production obtained with satellite-based estimates of whitecap fraction at frequency 37 GHz (equation 2b).

4. Summary

Surface fluxes of CO_2 and sea spray production have been calculated using satellite-based data for whitecap fraction. The calculated values compare reasonably well with previous estimates of CO_2 transfer velocity (within 10%–60%) and sea spray mass flux (differences less than a factor of 2). Use of active whitecap fraction values would predict CO_2 transfer velocity and flux too low (approximately an order of magnitude) compared to direct observations. The results thus suggest that the upper limit of W_A

data combined with the lower limit *W* data would provide whitecap fraction and bubble-mediated CO_2 transfer values that would reproduce the gas exchange characteristics observed with the in situ measurements. For sea spray production flux, satellite-based whitecap fraction at both 10 and 37 GHz produce fair comparisons with previous estimates. A notable implication of these results is the possibility to use satellite-based radiometric measurements of whitecap fraction to obtain surface sea spray production and bubble-mediated gas transfer on a global scale.

Acknowledgments
This paper is dedicated to the memory of Ed Andreas, a long-admired colleague, mentor (as a member of my PhD committee), and friend. Ed's deep knowledge of physical oceanography and boundary layer turbulence, innovative research on sea spray production, vast field experience in the Arctic and Antarctic, and mastery of technical writing have shaped my work and career for years. In friendly conversations and regular correspondence, Ed was invariably inspiring and generous with his advice, trust, and encouragement. He is greatly missed.

References
[1] Wang Q *et al* 1995 *Air-Water Gas Transfer* ed B Jähne and E C Monahan (Hanau: AEON Verlag & Studio) 217
[2] Anguelova M D and Webster F 2006 *J. Geophys. Res.* **111**, C03017
[3] Gasier P W *et al* 2004 *IEEE Trans. Geosci. Remote Sens*, **42** 2347
Bettenhausen M *et al* 2006 *IEEE Trans. Geosci. Remote Sens.* **44** 597
Anguelova M D and Gaiser P W 2013 *Rem. Sen. Environ.* **139** 8196
[4] Anguelova M D *et al* 2010 *Proc. 17th Air-Sea Interaction Conf.* 174036
[5] Salisbury D J *et al* 2013 *J. Geophys. Res.-Oceans* **118** 6201
Salisbury D J *et al* 2014 *Geophys. Res. Lett.* **41** 1616
[6] Anguelova M D and Gaiser P W 2011 *J. Geophys Res.* **116** C11002
[7] Mellville W K 1996, *Annu. Rev. Fluid Mech.* **28**,279
Asher W *et al* 1998 *J. Mar. Tech. Soc.* **32** 32
[8] Anguelova M D and Hwang P A 2012 *Proc. 18th Air-Sea Interaction Conf.* 209178
[9] Phillips O M 1985 *J. Fluid Mech.* **156** 505
[10] Anguelova M D and Hwang P A 2015, *J. Phys. Oceanogr.*, doi: 10.1175/JPO-D-15-0069.1
[11] Fairall C W *et al* 1996 *J. Geophys. Res.*,**101** 3747
Fairall C W *et al* 2003 *J. Clim.* **16** 571
[12] Fairall C W *et al* 2011 *J. Geophys. Res.* **116** C00F09
[13] Woolf D K 1997 *Surface and Global Change* ed P S Liss and R A Duce (Cambridge, UK: Cambridge Univ. Press) 173
[14] Edson J B *et al* 2011 *J. Geophys. Res.* **116** C00F10
[15] Ho D T *et al* 2011 *J. Geophys. Res.* **116** C00F08
[16] Monahan E C and O'Muircheartaigh I G 1980 *J. Phys. Oceanogr.* **10**, 2094
Ho D T *et al* 2006 *Geophys. Res. Lett.* **33** L16611
[17] Andreas E L 2002, *Atmosphere-Ocean Interactions, Adv. Fluid Mech.* **33**, ed W Perrie (Boston: WIT) 1–46
[18] Monahan E C *et al* 1986 *Oceanic Whitecaps and Their Role in Air-Sea Exchange Processes* ed E C Monahan and G Mac Niocaill (Dordrecht, The Netherlands: Reidel Publishing Company) 167-174
[19] Andreas E L 1992 *J. Geophys. Res.* **97** 11429
[20] Albert M F M A *et al* 2015 *Atmos. Chem. Phys. Discuss.* **15**, 21219
[21] Jaeglé L *et al* 2011 *Atmos. Chem. Phys.* **11** 3137
[22] Sofiev M *et al* 2011 *J. Geophys. Res.* **116** D21302
[23] Norris S J *et al* 2008 *Atmos. Chem. Phys.* **8** 555
[24] Grythe H *et al* 2014 *Atmos. Chem. Phys.* **14** 1277

The potential role of sea spray droplets in facilitating air-sea gas transfer

E L Andreas[1], P Vlahos[2] and E C Monahan[2]

[1] NorthWest Research Associates, Inc., 25 Eagle Ridge, Lebanon, NH 03766, USA
[2] Department of Marine Sciences, University of Connecticut, 1080 Shennecossett Road, Groton, CT 06340, USA

E-mail: penny.vlahos@uconn.edu

Abstract. For over 30 years, air-sea interaction specialists have been evaluating and parameterizing the role of whitecap bubbles in air-sea gas exchange. To our knowledge, no one, however, has studied the mirror image process of whether sea spray droplets can facilitate air-sea gas exchange. We are therefore using theory, data analysis, and numerical modeling to quantify the role of spray on air-sea gas transfer. In this, our first formal work on this subject, we seek the rate-limiting step in spray-mediated gas transfer by evaluating the three time scales that govern the exchange: τ_{air}, which quantifies the rate of transfer between the atmospheric gas reservoir and the surface of the droplet; τ_{int}, which quantifies the exchange rate across the air-droplet interface; and τ_{aq}, which quantifies gas mixing within the aqueous solution droplet.

1. Introduction

Early in the study of air-sea gas transfer—roughly 35 years ago—ocean scientists began considering bubble-mediated transfer as a way to explain the nonlinear increase in the gas transfer velocity that becomes apparent when the 10-m wind speed (U_{10}) reaches about 10 m s^{-1} (Thorpe 1982; Merlivat and Memery 1983; Monahan and Spillane 1984). After all, stage-B whitecaps, which are the source of most ocean bubbles, reach a fractional ocean coverage of 1% at U_{10} = 10 m s^{-1}, and that coverage increases faster than the cube of the wind speed (Monahan and O'Muircheartaigh 1980). In the intervening years, many others took up the study of bubble-mediated gas transfer (e.g., Woolf and Thorpe 1991; Wallace and Wirick 1992; Keeling 1993; Woolf 1993, 2005; Asher *et al.* 1996; Monahan 2002; Vlahos and Monahan 2009; Vlahos *et al.* 2011).

Surprisingly few, however, have considered what appears to be the mirror-image process of bubble-mediated air-sea gas transfer—spray-mediated transfer. We therefore begin here a formal treatment of how and whether sea spray might influence the rate of air-sea gas transfer. We will generally follow the approach that Andreas and colleagues used for assessing how sea spray affects air-sea heat and moisture transfer (Andreas 1992, 2010; Andreas and DeCosmo 2002; Andreas *et al.* 1995, 2015). Microphysical modeling of the temperature and radius evolution of individual spray droplets underlies most of this work. In turn, we will begin our study of spray-mediated gas transfer by first studying the microphysical, thermodynamic, and gas transfer properties of individual spray droplets. We quantify these processes with time scales that ultimately reveal where the rate-limiting step is for spray-mediated gas transfer and, consequently, inform our decision on how next to proceed.

Content from this work may be used under the terms of the Creative Commons Attribution 3.0 licence. Any further distribution of this work must maintain attribution to the author(s) and the title of the work, journal citation and DOI.
Published under licence by IOP Publishing Ltd

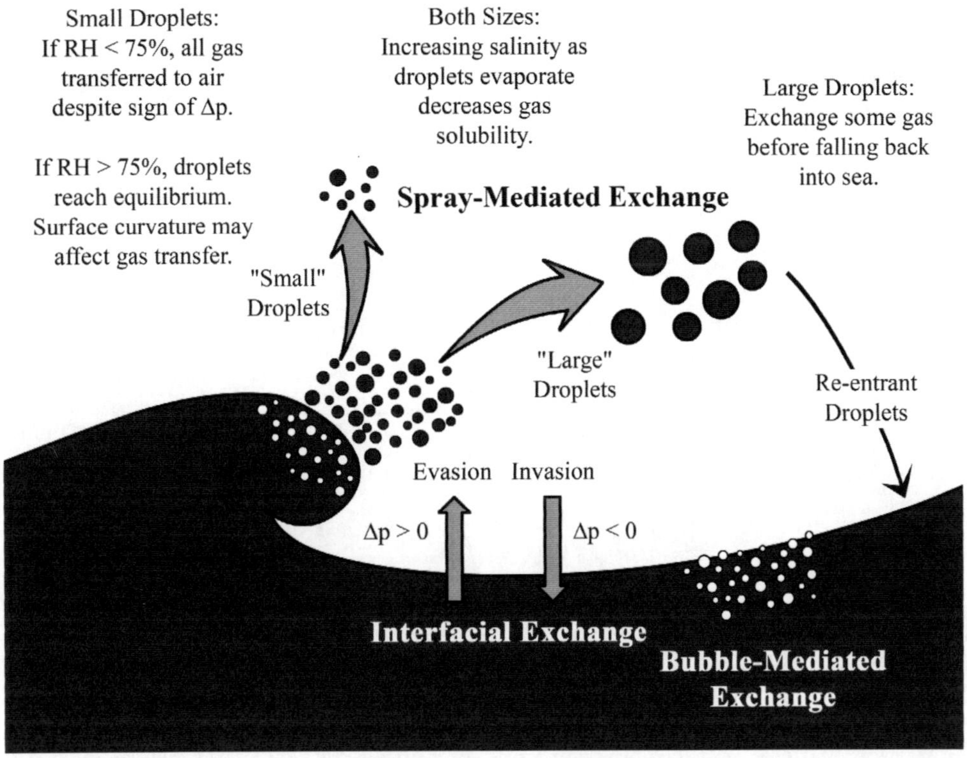

Figure 1. Gas transfer processes near the air-sea interface. The figure depicts three routes for gas transfer: interfacial exchange, which follows the sea-air difference in the partial pressure of the gas, Δp; bubble-mediated exchange; and spray-mediated exchange, in which droplet size is a crucial parameter.

Figure 1 is a schematic overview of air-sea gas transfer. The air and sea are always exchanging gases across the interface as a consequence of Henry's law equilibrium and the differences in partial pressures of the gases in the sea and in the near-surface air (Δp). This *interfacial* exchange goes approximately linearly with wind speed (Jähne *et al.* 1987; Woolf 2005). The copious bubbles that begin to form once whitecapping begins when the 10-m wind speed reaches about 4 m s^{-1} provide additional contact between air and sea and affect the gas transfer rate. Because whitecap coverage increases approximately with the cube of the wind speed, *bubble-mediated* gas transfer is presumed to go roughly as the cube of the wind speed (Monahan and Spillane 1984; Monahan 2002; Woolf 2005).

Also coincident with the onset of whitecapping is sea spray production. As with bubbles, spray is intimately tied to wave breaking; the number of spray droplets produced therefore also increases as roughly the third power of wind speed. In most marine environments, spray droplets cool rapidly to a temperature lower than the air thermometer. They also evaporate slowly, thereby losing fresh water and becoming more saline.

Figure 1 identifies two classes of spray droplets: small and large. The distinction is relative, not absolute, as we will explain later. Small droplets remain suspended long enough to come to radius (evaporative) equilibrium with the environment; large droplets fall back into the sea before reaching radius equilibrium—these are *re-entrant* spray droplets.

The small droplets generally transport gas from the ocean to the atmosphere because they lose fresh water, become more saline, and therefore become an increasingly inhospitable environment for dissolved gases. If the near-surface relative humidity is 75% or greater, the efflorescence point for saline droplets, these droplets reach equilibrium as aqueous solution droplets. In an environment with relative humidity less than 75%, these small droplets ultimately lose all their water and all their gas to the atmosphere and leave behind just a dry salt crystal.

While large droplets carry most of the mass in the spray, they cannot participate fully in air-sea gas transfer because of limits on their lifetime. They return to the ocean before they become hypersaline and, thus, can exchange gas either to or from the ocean depending, mostly, on the sign of Δp. The time scales that govern the rates for these various processes are our subject here.

2. Theoretical background

2.1. Spray droplet evolution

In the early stages of evolution, both droplet temperature (T) and radius (r) approximately follow exponential relations:

$$\frac{T(t) - T_{eq}}{T_w - T_{eq}} = \exp(-t/\tau_T), \tag{1}$$

$$\frac{r(t) - r_{eq}}{r_0 - r_{eq}} = \exp(-t/\tau_r). \tag{2}$$

Here, t is the time since the droplet formed; T_w, the sea surface temperature (the initial droplet temperature); r_0, the initial droplet radius; T_{eq} and r_{eq}, the droplet's equilibrium temperature and radius; and τ_T and τ_r, the corresponding e-folding times to reach these equilibrium values. Later figures show values of τ_T and τ_r. The microphysical quantities T_{eq}, r_{eq}, τ_T, and τ_r all depend on the initial droplet radius, temperature, and salinity and on environmental conditions like air temperature, relative humidity, and barometric pressure.

2.2. Gas time scales

In the context of the time scales τ_T and τ_r, over which spray droplets themselves evolve, we turn next to consider the time scales that govern how gas molecules interact with water droplets. The cloud physics literature—namely, the books by Seinfeld and Pandis (2006), Pruppacher and Klett (2010), and Lamb and Verlinde (2011) — have provided the background for this work.

Figure 2 schematically shows how an individual spray droplet may exchange an arbitrary gas with an atmospheric reservoir. Although spray-mediated transfer can involve both invasion and evasion, we suspect evasion will be the primary direction for spray-mediated air-sea gas transfer.

The Henry's law coefficient (K_H) is crucial in all studies of gas transfer across an air-water interface. We define K_H for spray droplet geometry according to the relation

$$C_d(r_0) = K_H\, p_g(r_0). \tag{3}$$

Here, $p_g(r_0)$ is the partial pressure (or fugacity) of arbitrary gas g at the external surface of a droplet, and $C_d(r_0)$ is the concentration of the gas just inside the droplet (see figure 2). Our preferred units for p_g and C_d are atm and mol L^{-1}, respectively. Hence, K_H has units of mol L^{-1} atm^{-1}.

To put all gas quantities in (3) in terms of a concentration, we use the ideal gas law to convert p_g to the concentration of the gas in air (C_a). Equation (3) thus becomes

$$C_d(r_0) = K_H\, RT\, C_a(r_0), \tag{4}$$

where C_a also has units of mol L−1. Also in (4), R (= 8.20574 x 10^{-2} atm L mol^{-1} K^{-1}) is the universal gas constant, and T is the kelvin temperature. Notice, the product *KHRT* is dimensionless. Equation (4) explains the apparent discontinuity in concentration in figure 2 at the surface of the spray droplet.

All three of our primary sources — Seinfeld and Pandis (2006, pp. 556–557), Pruppacher and Klett (2010, p. 764), and Lamb and Verlinde (2011, p. 508) — estimate the time scale τ_{aq} that quantifies mixing inside the aqueous solution droplet (figure 2) as

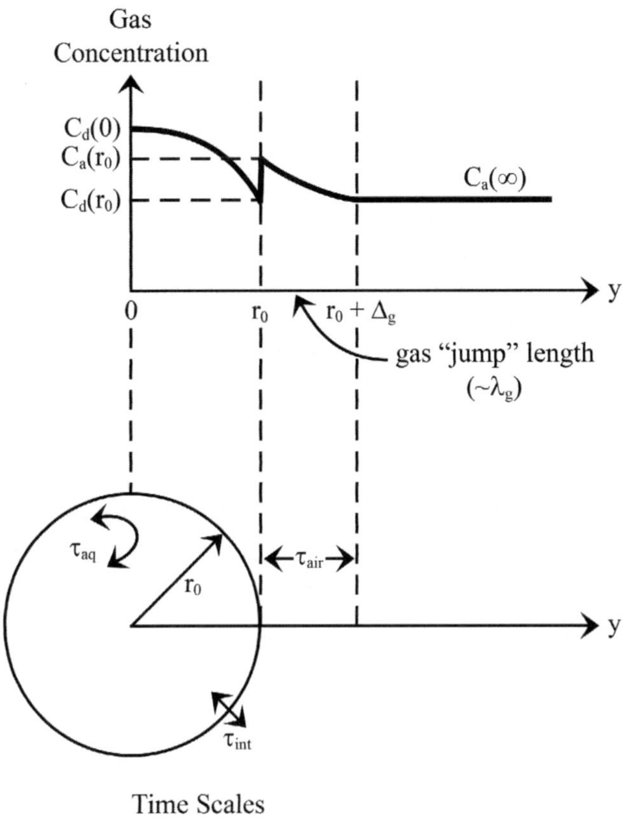

Figure 2. A gas evasion example. The upper part of the figure shows the gas concentration C as a function of radial distance y from the center of a spray droplet. The apparent discontinuity in concentration at r_0 is required by Henry's law equilibrium. The lower part depicts the three time scales (τ_{air}, τ_{int}, τ_{aq}) relevant to spray-mediated gas exchange. Adapted from Seinfeld and Pandis (2006, p. 550) and Lamb and Verlinde (2011, p. 502).

$$\tau_{aq} = \frac{r_0^2}{\pi^2 D_{g,sw}}, \tag{5}$$

where $D_{g,sw}$ is the molecular diffusivity of gas g in seawater.

Using an approach outlined by Pruppacher and Klett (2010, p. 761), we derive estimates for τ_{air} and τ_{int} of

$$\tau_{air,PK} = \frac{r_0^3 K_H R T_d}{3 D_g \left(r_0 + \Delta_g\right)}, \tag{6}$$

and

$$\tau_{int,PK} = \frac{r_0 K_H R T_d}{3\alpha}\left(\frac{2\pi M_g}{R T_a}\right)^{1/2}. \tag{7}$$

In these, D_g is the molecular diffusivity of the gas in air; M_g the molecular weight of the gas; T_a the air temperature; and T_d the droplet temperature (assumed uniform in this derivation). If the droplet is small enough, the air and gas may no longer behave as continuous fluids; the Δ_g in (6), the gas "jump" length, which is approximately the mean free path of the gas in air (λ_g; see figure 2), determines what "small enough" is. Gas molecules cross this jump length as dictated by the kinetic theory of gases (e.g., Bohren and Albrecht 1998, pp. 188, 234–235). Thus, in (7), we recognize $(2\pi M_g/RT_a)^{1/2}$ as $4/\overline{v}_g$, where $\overline{v}_g$ is the average speed of an ideal gas molecule according to the Maxwell–Boltzmann speed distribution (Bohren and Albrecht 1998, pp. 60–64).

Finally, in (7), α is the mass accommodation coefficient of the gas: the ratio of the number of gas molecules that stick to a droplet over the number that strike the droplet. The values of α for our six gases are still emerging, particularly in the case of sea spray where the evaporative flux will require consideration of significant changes in the efflux rates as the drop shrinks. In a comparable microphysical model for heat and water vapor exchange with a spray droplet, Andreas (1989, 1990) used $\alpha_v = 0.036$ for water vapor (Pruppacher and Klett 2010, table 13-1). Based on Davis (2006), values of unity are appropriate for condensation coefficients at relatively low temperatures such that re-emission of the gas does not occur, but the emerging accommodation coefficient (what leaves the droplet) is the most appropriate scaler for a spray model. It is important to note that values of unity are based on the assumption that the Knudsen number is sufficiently large, though this may not hold in a shrinking droplet (Davis 2006). Also, values based on isolated droplets tend to overestimate evaporation rates by at least 65% (Devarakonda *et al.* 1998) avoiding attenuation due to droplet interactions. In the absence of compelling evidence for a definitive choice, we conservatively apply both 0.036 and 1.0 across all gases and evaluate the sensitivity of this parameter in the interpretation of our results.

Although Lamb and Verlinde (2011, p. 507) derive the same estimate as (7) for the interfacial time scale (i.e., $\tau_{int,LV} = \tau_{int,PK}$), they ignore the jump layer and, thereby, derive as the time scale in air [cf. (6)]

$$\tau_{air,LV} = \frac{r_0^2 K_H R T_d}{3 D_g}. \tag{8}$$

In contrast with (6) and (8), Seinfeld and Pandis (2006, pp. 538, 549–551, 580–582) assume that the gas concentration at the external surface of the spray droplet is constant with time and thereby estimate the gas time scale in air as

$$\tau_{air,SP} = \frac{r_0^2}{4 D_g}. \tag{9}$$

But they do a much more interesting derivation of the interfacial time scale, τ_{int}.

On allowing the gas concentration in the droplet to vary with radial position, Seinfeld and Pandis (2006, pp. 551–554) derive

$$\tau_{int,SP} = \frac{r_0^2}{\beta_1^2 D_{g,sw}}, \tag{10}$$

where β_1 is the first positive root of

$$\beta_n \cot(\beta_n) + L = 0. \tag{11}$$

In (11),

$$L = \frac{r_0 H}{D_{g,sw}} - 1, \tag{12}$$

where

$$H = \frac{\alpha}{K_H \left(2\pi M_g R T_a\right)^{1/2}}. \tag{13}$$

For large values of L, nominally above 100, β_1 from (11) is π (e.g., Carslaw and Jaeger 1996, p. 492) and

$$\lim_{L \to \infty} \tau_{\text{int},SP} = \frac{r_0^2}{\pi^2 D_{g,sw}}. \tag{14}$$

We will see this case often in our subsequent calculations because L is large when r_0 is large, α is large, or the gas is not very soluble (i.e., small K_H). Of our six gases, all except carbon dioxide would be termed only weakly soluble. Notice also that (5) and (14) are identical.

Equations (5) and (10) follow from the assumption that the fluid within the droplets is motionless; gas molecules thus would move only through molecular diffusion, $D_{g,sw}$. Ample evidence exists, however, that the fluid, even in small droplets, develops circulations when a shear stress occurs at the droplet's surface (Clift *et al.* 1978, pp. 36–38, 127–129; Pruppacher and Klett 2010, pp. 386–393). Consequently, mixing of the gas within a droplet and across its interface is surely much faster than τ_{aq} and $\tau_{int,SP}$ suggest. We thus consider (5) and (10) as upper bounds for these time scales.

2.3. Droplet residence time

A droplet's residence time in air provides the context for all of these time scales. In short, a spray droplet can exchange heat, water vapor, and gases with air only during its (often brief) lifetime between creation and re-entry into the ocean.

Because spume droplets, which are created by the wind's tearing droplets right off the wave crests, accomplish most of the spray-mediated heat and moisture exchange, Andreas and colleagues (Andreas 1992; Andreas and DeCosmo 2002; Andreas *et al.* 2015) based a residence time scale on the behavior of these droplets:

$$\tau_f = \frac{H_{1/3}/2}{u_f(r_0)}. \tag{15}$$

Here, $u_f(r_0)$ is the terminal fall velocity of droplets of radius r_0. Also, $H_{1/3}$ is the significant wave height; consequently, $H_{1/3}/2$ ($\equiv A_{1/3}$) is the significant wave amplitude, the height above mean sea level where many spray, and most spume, droplets originate. Equation (15), in essence, models the residence time in air of a droplet undergoing ballistic flight.

3. Spray droplet time scales

Figure 3 compares these several time scales over a range of radii for one set of environmental conditions and for two gases: helium and carbon dioxide. In both plots, the τ_f, τ_r, and τ_T curves are the same because these times do not depend on the gas.

We selected helium (M_g = 4 g mol^{-1}) and carbon dioxide (M_g = 44 g mol^{-1}) for display in figure 3 because their molecular masses span the range from smallest to largest in our set of six gases and thereby demonstrate how the size of the gas molecule influences τ_{air}, τ_{int}, and τ_{aq}.

This is how we interpret figure 3. Any spray-mediated transfer must occur in a time less than the droplet's residence time, τ_f. This time scale decreases as the radius increases because u_f increases with radius. The τ_f curves in figure 3 are for a 10-m wind speed of 12 m s^{-1}; the τ_f curve will rise for increasing wind speed (because $H_{1/3}$ increases), allowing more spray-mediated exchange to occur. The τ_f curve in figure 3, as an example, is above τ_r for radii up to about 30 μm. Droplets of 30 μm and smaller thus can reach radius equilibrium and become more saline before they fall back into the ocean. Droplets larger than 30 μm, where $\tau_f < \tau_r$, on the other hand, fall back into the water before losing much water and, consequently, return to the ocean with approximately their original salinity. A droplet's temperature evolution is much faster. Figure 3 shows that droplets up to about 200 μm in radius have time to reach an equilibrium temperature of T_{eq} at this wind speed before they fall back into the sea.

Our several estimates of gas time scales are comparable to and even shorter than τ_T. At first sight, we might say mixing of gas molecules within a spray droplet—parameterized as τ_{aq}—is the slow step in spray-mediated gas transfer because, at least in these two figures, it is the largest of the gas time scales. (In figure 3, $\tau_{aq} = \tau_{int,SP}$. More on this shortly.) But remember, τ_{aq} is a maximum value that does not recognize any fluid motion in a spray droplet that would certainly enhance gas mixing by at least an order of magnitude. Consequently, we suspect that the true τ_{aq} will be at least ten times shorter than depicted.

For the air time scales, τ_{air}, in figure 3, we show only $\tau_{air,PK}$ (6), because it is near the Lamb and Verlinde (2011) scale (8) and because the Seinfeld and Pandis (2006) scale (9) is based (unrealistically) only on molecular diffusion of gas molecules to and from a spray droplet. Notice in figure 3 that $\tau_{air,PK}$ is orders of magnitude less than τ_T, which we consider to represent very fast exchange. $\tau_{air,PK}$ does, however, increase by roughly three orders of magnitude between helium and carbon dioxide as a consequence of how increasing molecular weight slows molecular velocities in air.

The final time scale shown in figure 3 quantifies exchange across a droplet's interface, and we represent this exchange with both $\tau_{int,PK}$, (7), and $\tau_{int,SP}$, (10). For helium and carbon dioxide and our choice of the accommodation coefficient, 0.036, $\tau_{int,SP}$ has the same calculated value as τ_{aq}—compare (5) and (14). For an accommodation coefficient of 1 the $\tau_{int,SP}$ becomes 2 orders of magnitude smaller. And because $\tau_{int,SP}$ is formulated in terms of the molecular diffusivity in the spray droplet, $D_{g,sw}$, we expect it to be lower, too, if we acknowledge the likelihood of fluid motion within spray droplets. The second interfacial time scale, $\tau_{int,PK}$, (7), is orders of magnitude less than the Seinfeld and Pandis (2006) prediction, $\tau_{int,SP}$. A mass accommodation coefficient of unity further reduces the $\tau_{int,PK}$ by a factor of 30. And although $\tau_{int,PK}$ does increase with the increasing molecular weight of the gas molecule, it always represents extremely fast gas transfer over the chosen mass accommodation coefficient range.

4. Conclusions

Gas molecules transit between the deep interior of a sea spray droplet and an atmospheric reservoir in five distinct steps. 1) The molecules must diffuse between the deep interior of the droplet and its interior surface (time scale τ_{aq}), 2) must cross the air-droplet interface (time scale τ_{int}), and 3) the molecules must move between the ambient atmosphere and the exterior surface of the droplet (time scale τ_{air}). Within spray droplets, 4) some gas molecules will dissociate, and 5) some will undergo chemical reactions.

Of the six gases we are currently considering, all must complete the first three steps for spray-mediated transfer to occur; but only carbon dioxide participates in the additional fourth and fifth steps—dissociation and chemical reactions. Hence, for helium, neon, argon, oxygen, and nitrogen, the time scales we have evaluated are sufficient for modeling their role in spray-mediated transfer. In future work, we will consider how the chemical behavior of carbon dioxide influences the rate at which it undergoes spray-mediated exchange.

We found several opinions as to how to parameterize the time scales for transfer in the air, τ_{air}, and transfer across the air-droplet interface, τ_{int}. For τ_{air}, we view the time scale based on Pruppacher and Klett's (2010) approach ($\tau_{air,PK}$) as most appropriate because it is similar to Lamb and Verlinde's (2011) result and more physical that Seinfeld and Pandis's (2006). In contrast, the interfacial time scale from Seinfeld and Pandis ($\tau_{int,SP}$) seems more appropriate than Pruppacher and Klett's because it more realistically treats gas diffusion within spray droplets. Meanwhile, all our main sources agree on the aqueous time scale, τ_{aq}, that parameterizes diffusive mixing within droplets.

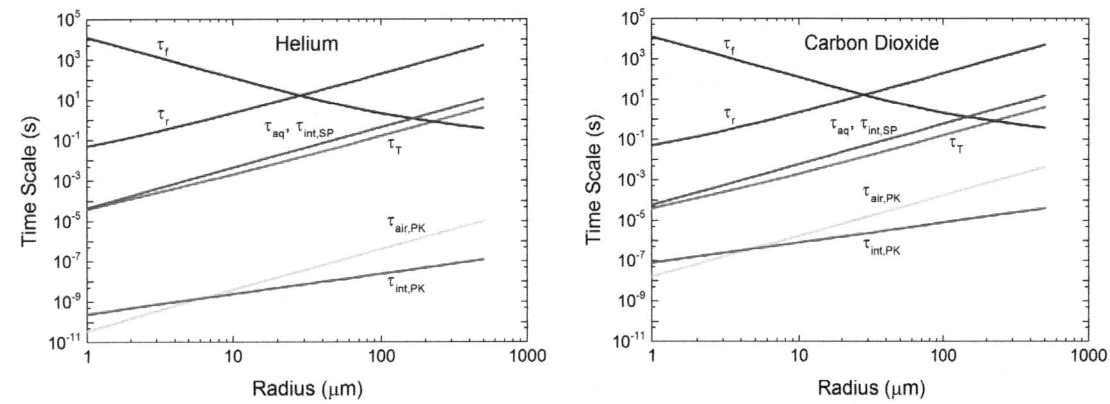

Figure 3. For droplets of radius r_0, the figure shows the time scales for temperature (τ_T) and radius (τ_r) evolution, the droplet residence time (τ_f), and the time scales that govern transfer of helium (left panel) and carbon dioxide (right panel) molecules to or from a spray droplet in air (τ_{air}), across the air-droplet interface (τ_{int}) and within the droplet (τ_{aq}). The interfacial time scale derives from both Pruppacher and Klett (2010; $\tau_{int,PK}$), equation (7), and Seinfeld and Pandis (2006; $\tau_{int,SP}$), equation (10). The time scale in air derives from Pruppacher and Klett ($\tau_{air,PK}$), equation (6); the aqueous mixing time scale (τ_{aq}) is (5). Conditions here are for a 10-m wind speed of 12 m s^{-1}, air temperature and water temperature of 20°C, relative humidity of 85%, and sea surface salinity of 34 psu; the mass accommodation coefficient is set at $\alpha = 0.036$. For this value, β_1 in the Seinfeld and Pandis $\tau_{int,SP}$ is π, and $\tau_{int,SP}$ is the same as τ_{aq}.

A caveat regarding $\tau_{int,SP}$ and τ_{aq} is that mixing within spray droplets likely also results from fluid motions, not just from molecular processes quantified by the molecular diffusivity of a gas in seawater, $D_{g,sw}$. In light of this physics, both exchange across the droplet interface and within a droplet could be one or more orders of magnitude faster than estimated from $\tau_{int,SP}$ and τ_{aq}.

As a consequence, τ_{air} and our revised estimates of $\tau_{int,SP}$ (regardless of mass accommodation coefficient values 0.036 or 1.0) and τ_{aq} are all shorter than the microphysical time scales τ_T and τ_r that quantify a spray droplet's physical evolution. Hence, to estimate spray-mediated gas transfer, we can reasonably assume that the gas in a spray droplet is in instantaneous equilibrium with the atmospheric reservoir. The gas concentration in the droplet will therefore evolve as the radius, temperature, and salinity of the droplet evolve. Thus our next step is to track simultaneously this gas concentration for a spectrum of droplets as they are injected into the near-surface marine atmosphere and evolve according to the Andreas (1989, 1990) microphysical model.

Acknowledgments
This paper is dedicated to our late esteemed colleague Dr. Edgar L. Andreas, whose pioneering work in sea spray continues to inspire. We thank Stephen Schwartz and Bill Asher for discussion on the mass accommodation coefficient and Emily Moynihan of Blythe Visual for preparing figures 1 and 2. The U.S. National Science Foundation supported this work with awards 13-56377 (ELA) and 13-56541 (PV and ECM).

References
[1] Andreas E L 1989 *CRREL Rep. 89-11*, (Hanover, NH: US Army Cold Regions Research and Engineering Laboratory)
[2] Andreas E L 1990 *Tellus* **42** 481
[3] Andreas E L 1992 *J. Geophys. Res.* **97** 11429
[4] Andreas E L 2010 *J. Phys. Oceanogr.* **40** 608
[5] Andreas E L and DeCosmo J 2002 *Bound.-Layer Meteor.* **103** 303

[6] Andreas E L *et al* 1995 *Bound.-Layer Meteor.* **72** 3
[7] Andreas E L *et al* 2015 *Quart. J. Roy. Meteor. Soc.* **141** 642
[8] Asher W E *et al* 1996 *J. Geophys. Res.* **101** 12027
[9] Bohren C F and Albrecht B A 1998 *Atmospheric Thermodynamics* (New York: Oxford University Press)
[10] Carslaw H S and Jaeger J C 1996 *Conduction of Heat in Solids, Second Ed* (Oxford: Oxford University Press)
[11] Clift R *et al* 1978 *Bubbles, Drops, and Particles* (Mineola, NY: Dover)
[12] Jähne B *et al* 1987 *J. Geophys. Res.* **92** 1937
[13] Keeling R F 1993 *J. Mar. Res.* **51** 237
[14] Lamb D and Verlinde J 2011 *Physics and Chemistry of Clouds* (Cambridge, UK: Cambridge University Press)
[15] Merlivat L and Memery L 1983 *J. Geophys. Res.* **88** 707
[16] Monahan E C 2002 *Gas Transfer at Water Surfaces*, ed M A Donelan *et al* (Washington, DC: American Geophysical Union) 193–197
[17] Monahan E C and O'Muircheartaigh I 1980 *J. Phys. Oceanogr.* **10** 2094
[18] Monahan E C and Spillane M C 1984 *Gas Transfer at Water Surfaces*, ed W Brutsaert and G J Jirka (Dordrecht: D Reidel) 495–503
[19] Pruppacher H R and Klett J D 2010 *Microphysics of Clouds and Precipitation, 2nd Revised Ed.* (Dordrecht: Springer)
[20] Seinfeld J H and Pandis S N 2006 *Atmospheric Chemistry and Physics: From Air Pollution to Climate Change, Second Ed.* (Hoboken, NJ: John Wiley and Sons)
[21] Thorpe S A 1982 *Phil. Trans. Roy. Soc. London* A **304** 155
[22] Vlahos P and Monahan E C 2009 *Geophys. Res. Lett.* **36** doi:10.1029/2009GL040695
[23] Vlahos P *et al* 2011 *Gas Transfer at Water Surfaces 2010*, ed S Komori *et al* (Kyoto: Kyoto University Press) 455–463
[24] Wallace D W R and Wirick C D 1992 *Nature* **356** 694
[25] Woolf D K 1993 *Atmos.-Ocean* **31** 517
[26] Woolf D K 2005 *Tellus* B **57** 87
[27] Woolf D K and Thorpe S A 1991 *J. Mar. Res.* **49** 435

IOP Conf. Series: Earth and Environmental Science **35** (2016) 012004 doi:10.1088/1755-1315/35/1/012004

Breaking waves and near-surface sea spray aerosol dependence on changing winds: Wave breaking efficiency and bubble-related air-sea interaction processes

P A Hwang, I B Savelyev and M D Anguelova

Naval Research Laboratory, 4555 Overlook Ave SW, Washington DC 20375 USA

E-mail: paul.hwang@nrl.navy.mil

Abstract. Simultaneous measurements of sea spray aerosol (SSA), wind, wave, and microwave brightness temperature are obtained in the open ocean on-board Floating Instrument Platform (FLIP). These data are analysed to clarify the ocean surface processes important to SSA production. Parameters are formulated to represent surface processes with characteristic length scales spanning a broad range. The investigation reveals distinct differences of the SSA properties in rising winds and falling winds, with higher SSA volume in falling winds. Also, in closely related measurements of whitecap coverage, higher whitecap fraction as a function of wind speed is found in falling winds than in rising winds or in older seas than in younger seas. Similar trend is found in the short scale roughness reflected in the microwave brightness temperature data. In the research of length and velocity scales of breaking waves, it has been observed that the length scale of wave breaking is shorter in mixed seas than in wind seas. For example, source function analysis of short surface waves shows that the characteristic length scale of the dissipation function shifts toward higher wavenumber (shorter wavelength) in mixed seas than in wind seas. Similarly, results from feature tracking or Doppler analysis of microwave radar sea spikes, which are closely associated with breaking waves, show that the magnitude of the average breaking wave velocity is smaller in mixed seas than in wind seas. Furthermore, breaking waves are observed to possess geometric similarity. Applying the results of breaking wave analyses to the SSA and whitecap observations described above, it is suggestive that larger air cavities resulting from the longer breakers are entrained in rising high winds. The larger air cavities escape rapidly due to buoyancy before they can be fully broken down into small bubbles for the subsequent SSA production or whitecap manifestation. In contrast, in falling winds (with mixed seas more likely), the shorter breaker entrains smaller and more numerous air cavities that stay underwater longer for more efficient bubble breakup by turbulence and prolonging the surface disturbances attributable to wave breaking. For low winds, the breaking scale is small and with high efficiency for SSA or whitecap generation; the trend of rising or falling wind is less important.

1. Introduction

Breaking of ocean surface waves plays an important role in many air-sea interaction processes. For example, it is a main source of turbulence on the upper ocean layer; breaking waves entrain bubbles and enhance air-sea gas exchange; returning to the ocean surface, the entrained air bubbles appear as whitecaps enhancing the surface albedo and modifying heat and gas transfer; the returning bubbles also cause surface disturbances and contribute to the breaking-wave-induced surface roughness; the formation of bubbles detached from the entrained air cavity is the main source of underwater ambient

Content from this work may be used under the terms of the Creative Commons Attribution 3.0 licence. Any further distribution of this work must maintain attribution to the author(s) and the title of the work, journal citation and DOI.
Published under licence by IOP Publishing Ltd

noise; and bursting of bubbles at the sea surface is one of the main generation sources of sea spray aerosols, which represent an important member of the atmospheric aerosol population.

In this paper, we report results of a field experiment addressing the effect of wave breaking on SSA production. Simultaneous measurements of SSA, wind, wave and microwave brightness temperature are obtained in the open ocean on-board FLIP in the Pacific Ocean about 200 km off the California coast [1]. Over a period of about two weeks of free drifting, the experiment encountered a wide range of environmental conditions with wind speed U_{10} ranging between 1.9 and 15.6 m/s, and significant wave height H_s between 1.0 and 4.5 m. Based on the field notes, there was no rain event during the period reported in this paper. Here we focus on results showing distinctly higher SSA properties (number, volume, flux) in a falling wind event than that in a rising wind event (for the same wind speed). Similar results have been reported in Ovadnevaite *et al.* [2] and in the closely related whitecap observations of Callaghan *et al.* [3], though the physical reasons behind these observations are not clear.

In previous investigations of the length and velocity scales of breaking waves [4–8], it is found that the breaking wavelength (velocity) scale is shorter (slower) in mixed seas compared to that in wind seas. From the consideration of breaking wave geometric similarity [9], the size of the breaking patches and the air cavities entrained into water is proportional to the breaking wavelength. Due to the buoyancy effect, the large cavities escape soon after breaking entrainment. As a result, the shorter dwell time of the entrained large air cavity decreases the effective air volume useful for subsequent bubble cloud generation and other related processes. Thus, it is deduced that larger breaking is less efficient in bubble-related processes such as production of SSA or whitecaps. Considering that mixed seas are more likely to occur in falling wind events and wind seas in rising winds, we propose that the efficiency of wave breaking is the cause of the observed differences in the SSA generation or whitecap manifestation between rising and falling wind conditions. In other words, the breaking efficiency in these bubble-related processes is influenced by the breaking length scale through the entrained air cavity size, leading to the observed higher SSA or whitecaps in falling winds than in rising winds.

In the following, we present the analysis of the field measurements in section 2; the interpretation of the SSA differences in rising and falling winds is discussed in section 3 and a summary is given in section 4.

2. Analysis

The measured wind, wave and microwave brightness temperature are analyzed to formulate parameters that represent surface processes with distinctively different characteristic length scales ranging from a few centimeters to a few decameters. The short scale is represented by the roughness component of 10.7–GHz microwave brightness temperature ΔT_b. This quantity is derived from the difference between the horizontal and vertical polarizations of the wind-induced deviation of the brightness temperature from the flat surface values: $\Delta T_b = \delta T_{bH} - \delta T_{bV}$. Using the method described in Hwang [10], the wind-induced deviations δT_{bH} and δT_{bV} can be separated into the roughness and foam components, a brief summary is given in Appendix A. Because the foam components of δT_{bH} and δT_{bV} are small for the wind conditions in our experiment, they almost cancel each other out, leaving ΔT_b to be contributed predominantly by the centimeter scale surface roughness to the 10.7 GHz-emission.

The long-scale process is represented by the surface wave energy dissipation rate E_t. The wave energy dissipation rate is calculated from the energy conservation equation governing the wind-generated waves and coupled with the fetch- or duration-limited wind wave growth functions [11]. Expressed as a dimensionless function, it is given by $\frac{E_t}{\rho_a U_{10}^3} = 0.20\omega_*^{3.3}\eta_*$, where ρ_a is the density of air (1.2 kg/m^3), $\omega_* = \omega_p U_{10}/g$ and $\eta_* = \eta_{rms}^2 g^2/U_{10}^4$ are dimensionless wave spectral peak frequency ω_p and wave variance η_{rms}^2, respectively, U_{10} is the neural wind speed at 10-m elevation, and g is the gravitational acceleration. In mixed seas, it is necessary to separate the swell and wind sea components in order to make use of the energy dissipation rate equation derived from the wind wave growth

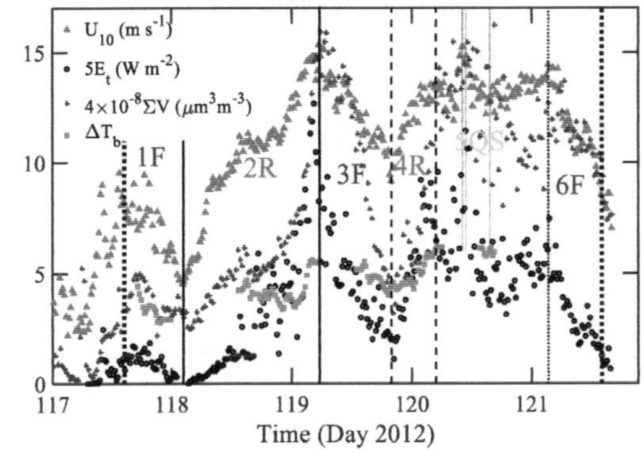

Figure 1. Time series of U_{10}, E_t, ΔT_b and ΣV, with six episodes (rising, falling, quasi-steady) identified.

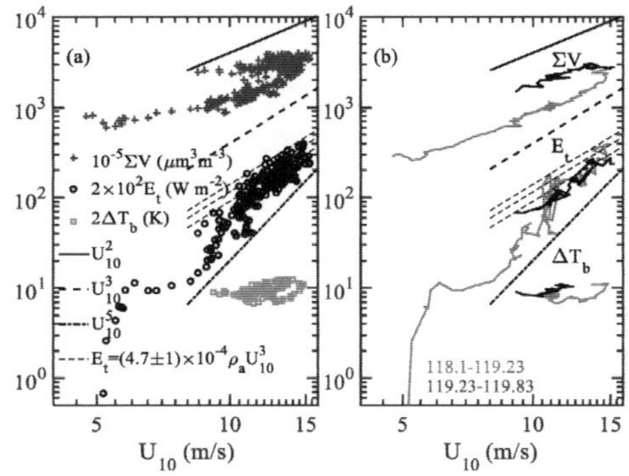

Figure 2. Power-law dependency of ΣV, E_t, and ΔT_b on U_{10}: (a) Data unsorted; and (b) 2 consecutive time segments of rising (red) and falling (black) wind events.

functions. The procedure of swell–sea separation described in Hwang *et al.* [12] is used in this study. Further detail is given in Appendix B.

Figure 1 shows the time series of wind speed U_{10}, energy dissipation rate E_t, brightness temperature deviation ΔT_b, and SSA volume ΣV, which represents the total atmospheric loading of aerosols with ambient radius ranging between 1 and 25 μm. The time series clearly illustrates the fluctuating feature of the events. ΣV, E_t, and ΔT_b show quadratic to cubic power-law dependence on wind speed for moderate to higher winds (figure 2a). For reference, U_{10}^2, U_{10}^3 and U_{10}^5 line segments are superimposed; they also serve as partitions separating various measurements. We are especially interested in the E_t parameter in this paper; also plotted are the set of empirical relations $E_t = (4.7 \pm 1)\times10^{-4}\rho_a U_{10}^3$ derived from examining the surface wave dissipation rates in several field datasets [11, 13]. The E_t dependence on wind speed in the present dataset is also close to cubic in higher winds.

A closer inspection reveals that the data are constituted of two main populations following similar wind speed dependence but with different absolute magnitudes. In figure 2b, the temporal evolutions of ΣV, E_t and ΔT_b in two consecutive time segments of rising and falling winds are highlighted by different colors (2R and 3F marked by vertical line segments in figure 1). Comparing the two events, a "hysteresis" in $\Sigma V(U_{10})$ is obvious. Figure 3 shows $\Sigma V(U_{10})$, $\Sigma V(E_t)$ and $E_t(U_{10})$ of all six events: three rising winds in black, two falling winds in cyan and one quasi-steady wind in green. The rising and falling wind events aggregate in separate clusters, and are the focus of this paper.

Because of the low settling velocity of aerosols, a tempting interpretation of the observed hysteresis is that once aerosols are suspended in rising winds, they stay aloft during a falling wind period thus resulting in higher number for a given wind speed in falling winds than in rising winds. However, a similar response lag is found in the roughness component of brightness temperature (figure 2b); the centimeter scale short surface waves contributing to the surface roughness certainly do not have the long dwell time of aerosol particles.

The contrast between falling and rising winds is conspicuously mixed in the wind sea energy dissipation rate in comparison to the SSA and ΔT_b observations. In the high wind region in particular, the energy dissipation rate is generally less in falling winds than in rising winds for the same wind speed. This is reasonable considering that rising winds carry additional forcing of wind speed acceleration. But then why the reverse trend in SSA flux and surface roughness? As noted in the last paragraph, although it is tempting to interpret the SSA observation of the hysteresis as due to the long dwell time of small aerosols once suspended in air, this argument cannot be applied to the short scale roughness as well as the whitecap observations, of which the constituent bubble lifetime is measured in seconds. A more plausible interpretation applicable to all three observations (SSA, whitecaps,

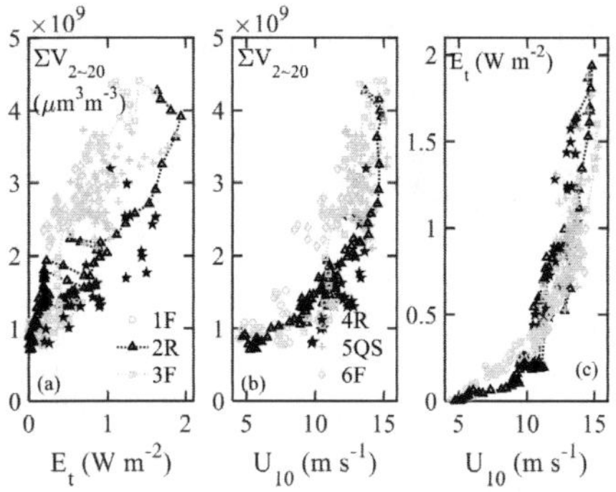

Figure 3. ΣV as a function of (a) E_t, and (b) U_{10}; and (c) $E_t(U_{10})$; rising winds (black), falling winds (cyan) and quasi-steady (green).

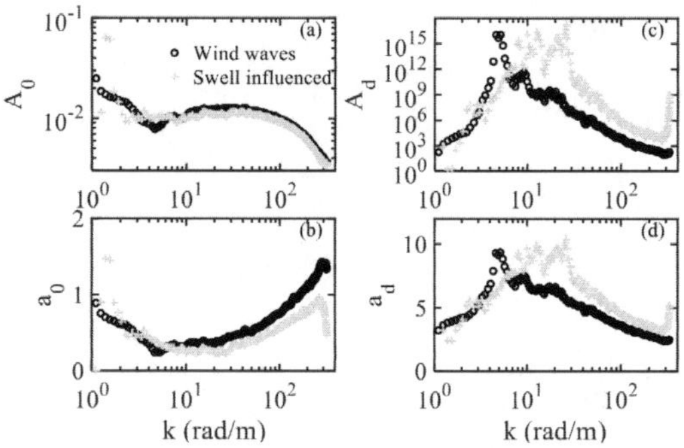

Figure 4. (a) A_0, and (b) a_0 of the roughness spectrum; (c) A_d, and (d) a_d of the dissipation function. Reproduced from [6]

brightness temperature deviation) is needed. This will be explored further with help from previous investigations of the length and velocity scales of breaking waves (section 3).

3. Discussion

In the research of length and velocity scales of breaking waves, field measurements show that the length scale of wave breaking is shorter in mixed seas than in wind seas. For example, following the analysis of source function balance described in Phillips [14], Hwang and Wang [6] present field data of short surface wave spectra obtained by a free-drifting wave gauge system. Experimental results led to the establishment of a similarity relation of the short wave spectral component: $B(u_*/c;k) = A_0(k)(u_*/c)^{a_0(k)}$, from which a general spectral dissipation function was formulated [14]: $f_d(B) = A_d(k)B^{a_d(k)}$. Sorting the field data into wind sea and mixed sea groups, the analysis shows that the characteristic length scale of the dissipation function shifts toward higher wavenumber (shorter wavelength) in mixed seas compared to that in wind seas, as illustrated in the upshift of the characteristic wavenumber of f_d from wind sea (black) to mixed sea (cyan) in figures 4c and 4d. Similarly, based on feature tracking [5] or Doppler analysis [7, 8] of microwave radar sea spikes, which are closely associated with breaking waves, the average breaking wave velocity is smaller with swell presence compared to that in pure wind seas (figure 5).

Here we clarify that the characteristic breaking length scale discussed in Hwang and Wang [6] is derived from the observations of wave spectral function variations with respect to the wind forcing parameter u_*/c. It therefore refers to the scale of waves that break, and not the length scale over which the wave energy dissipates. This distinction also applies to the observations of breaking velocities using feature tracking and Doppler analysis of sea spikes.

Applying the result of wave breaking analyses to the SSA, microwave brightness temperature and whitecap observations described in section 2, and coupled with the geometric similarities relating the breaking wavelength and breaking wave patch as well as the vertical and horizontal scales of breaking patches [9] as illustrated in figure 6, it is suggestive that larger air cavities resulting from the longer breakers are entrained in rising high winds. The larger air cavities escape rapidly due to buoyancy before they can be fully broken down into small bubbles for the subsequent SSA production or whitecap manifestation. In contrast, in falling winds (with mixed seas more likely), the shorter breaker entrains smaller and more numerous air cavities that stay underwater longer for more efficient bubble breakup by turbulence as well as prolonging the surface disturbance induced by wave breaking. For low winds, the breaking scale is small, also resulting in high efficiency of SSA or whitecap generation, and the trend of rising or falling in low winds is less distinguishable.

Based on our multisensor measurements, the E_t threshold of breaking efficiency transition in rising and falling winds is estimated to be 0.35 W/m^2 (figure 3a). As shown in figure 3b, the magnitude and

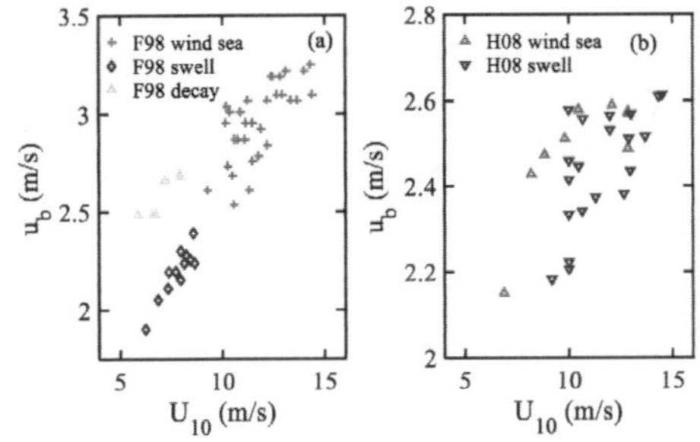

Figure 5. The breaking velocity deduced from (a) feature tracking [5] and (b) Doppler processing [7-8] of radar sea spikes. The breaking velocity is smaller in mixed seas (blue) in comparison to the value in wind seas (red). Adapted from figure 9 of Hwang *et al.* [7].

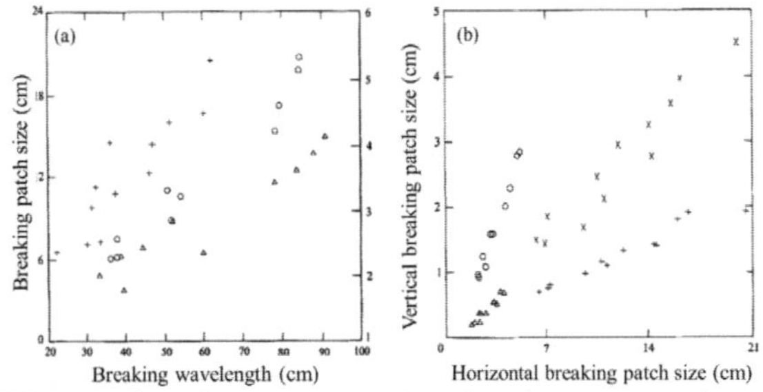

Figure 6. Laboratory results showing geometric similarity relating (a) the breaking wavelength and breaking patch size, and (b) the vertical and horizontal scales of breaking patches; different symbols represent different analyses. Adapted from figures 15 and 16 of Hwang *et al.* [9].

the rate of change of the SSA volume are considerably higher in falling winds (light symbols) than in rising winds (dark symbols). In comparison, the computed energy dissipation rate (figure 3c) shows mixed results: for high winds (>~10 m/s), the dissipation rate is slightly higher in rising winds than in falling winds; for low wind speeds (<~10 m/s) the trend is reversed. The higher rate of change represented by the slope of $\Sigma V(E_t)$ may be interpreted as a higher efficiency of breaking production in low or falling winds (figure 3a).

For a more comprehensive treatment, other physical aspects need to be considered. Examples include variation in wave growth rate in response to the nonstationarity of accelerating winds leading to an imbalance between dissipation and wind input in rising winds, and a potential relaxation of sea state towards equilibrium in falling winds leading to greater than average wave energy dissipation, essentially the opposite effect as in rising winds. Also, the relative size of air cavities entrained during falling and rising seas may be related to the spilling vs. plunging breakers. Photographic evidence suggests that plungers initially entrain larger air cavities than spillers [15]. Furthermore, it is necessary to investigate the timescales of air cavities escaping from underwater compared to the fragmentation time under turbulence [16]. These subjects, while interesting, cannot be addressed adequately due to space limitations.

As a final note, it is likely that the signature in the aerosol size spectra may be very similar between the dwell time consideration and the breaking efficiency consideration through the sizes of the breaking patches and entrained cavities. It is easy to deduce that if smaller particles are more abundant in falling winds than in rising winds, it could be evidence that particle residence time is responsible for increased concentrations during fallings winds. However, a more efficient breakup of the entrained air cavity also leads to the resulting bubble cloud shifting its population to smaller sizes, and the subsequent enhancement of small aerosols [17, 18].

Figure 7a shows four SSA volume spectra, each the average over 2 consecutive hours of measurements. These four spectra are selected, two each, in rising- and falling-wind periods. The average wind speeds are all within 11.0±0.1 m/s with standard deviations between 0.2 and 0.8 m/s. In general, the size range is broader in falling winds than in rising winds, consistent with our expectation from more efficient breakup of the entrained bubble cavities, but the increased small size population can also be from longer dwell time as discussed in the last paragraph. Figure 7b shows the SSA flux spectra and the comparison with three models [17, 18, 19]. The size distribution in our field data is close to exponential: $dF/d\ln r \sim \exp(-\alpha r)$. For r_{80}<~10 μm, the results are in best agreement, in both magnitude and size distribution, with the Monahan *et al.* bubble-based model [17], which may reflect the breaking efficiency mechanism of bubble-related processes as described in this paper. In summary, the breaking efficiency consideration provides a consistent explanation of the behavior of aerosols,

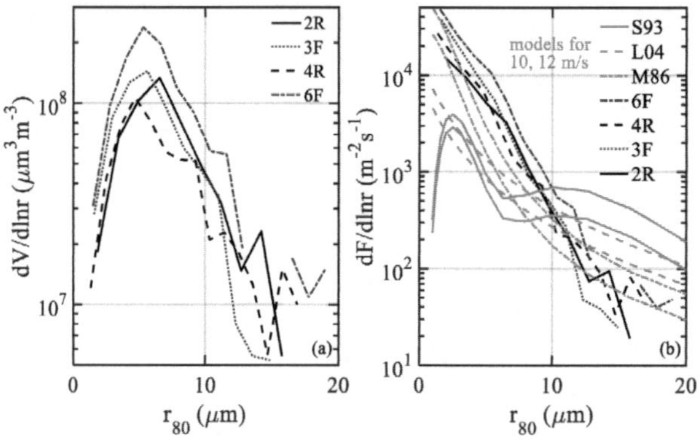

Figure 7. Four size spectra in rising and falling winds. The respective time period, mean RH and mean with standard deviation of U_{10} are [118.56~118.65, 119.64~119.73, 119.86~119.95, 121.45~121.53 d], [71.4, 81.7, 89.3, 84.6 %] and [10.9±0.4, 11.1±0.2, 11.0±0.8, 10.9±0.5 m/s] for [2R, 3F, 4R, 6F]): (a) volume spectra, and (b) flux spectra, also shown are computed results for U_{10}=10 and 12 m/s of three models [17, 18, 19], labelled respectively M86, L04 and S93.

whitecaps and brightness temperature in rising winds and falling winds. In comparison, the convenient argument of residence time can only be applied to the aerosol data but not the observations of whitecaps and brightness temperature.

For a more definitive analysis, it is necessary to have independent observations of the wave breaking properties, such as the entrained cavity size, depth of entrainment, the resulting bubble cloud size spectrum, and the turbulence properties. Our experiment does not have the necessary independent breaking observations for a definitive proof, so indirect evidence from previous relevant experiments and wave breaking analyses has been used. We hope that future observations can provide more direct measurements to support or refute the suggested hypothesis of wave breaking efficiency on bubble-related processes.

4. Summary

Several studies have shown distinct differences in the length and velocity scales of breaking waves in wind seas and mixed seas. It is deduced that the different breaking scales produce different sizes of breaking patches and air cavities entrained during the breaking process. This size difference may be the root cause of the observed higher SSA flux, whitecap coverage and surface roughness reflected in microwave brightness measurements in falling winds than those in rising winds.

Acknowledgments. This work is sponsored by the Office of Naval Research (NRL program elements 61153N). NRL publication number: NRL/PP/7260-15-0135. We are grateful for the comments and suggestions from an anonymous reviewer.

Appendix A: Foam and roughness components of microwave brightness temperature

There are two main components of the wind-induced deviation of the brightness temperature T_b from the flat surface value: foam and roughness. The formula connecting T_b and the sea surface temperature T_s can be written as [10]

$$T_{bp} = \left(e_{0p} + \delta e_{fp} + \delta e_{rp}\right)T_s, \tag{A1}$$

where subscript p is polarization, e_0 is the emissivity of the flat surface without foam, δe_f is the emissivity correction due to foam for the flat surface contribution, and δe_r is the rough surface contribution. These three components are given by:

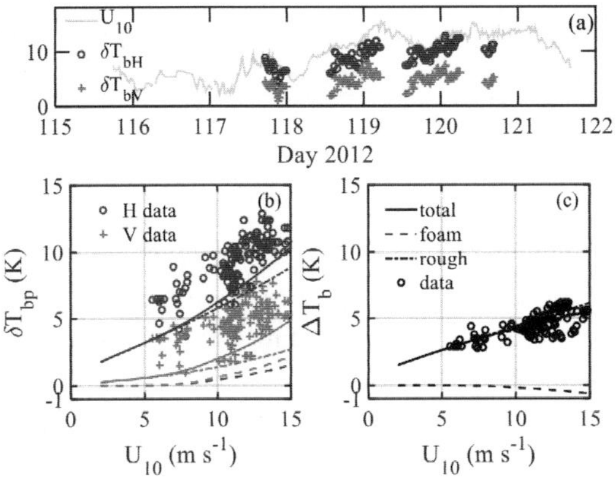

Figure A1. Processed results of the brightness temperature T_b measurements: (a) U_{10}, T_b and δT_{bp}. (c) $\Delta T_b=\delta T_{bH}-\delta T_{bV}$.

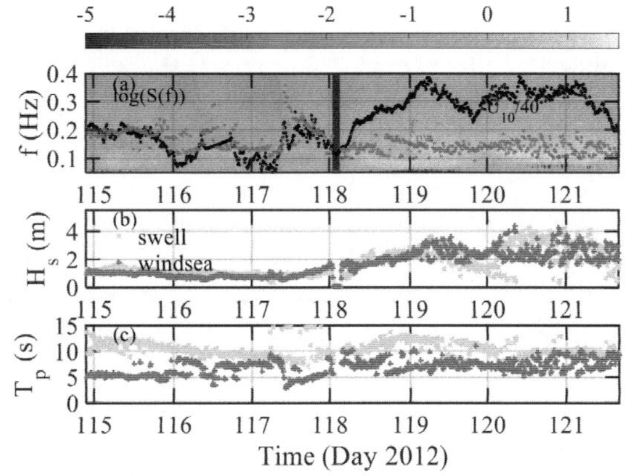

Figure B1. Sea state, swell and wind sea separation: (a) wave spectrum; (b) H_{sw} and H_{ss}; and (c) T_{pw} and T_{ps}.

$$e_{0p} = 1 - \left| R_{pp}^{(0)} \right|^2, \tag{A2}$$

where $R_{pp}^{(0)}(\varepsilon,\theta)$ is the Fresnel reflection coefficient of polarization p; the value is determined by the incidence angle θ and the frequency-dependent relative permittivity ε,

$$\delta e_{fp} = e_{ep} - e_{0p}, \tag{A3}$$

where e_{ep} is the effective emissivity of the flat surface with foam, and

$$\delta e_{rp} = \int_0^{\infty}\int_{-\pi}^{\pi} W(k',\phi') g_p(f,\theta,\phi,\varepsilon_e,k',\phi') k' d\phi' dk', \tag{A4}$$

where W is the directional ocean surface roughness spectrum with k the surface wave number and ϕ the propagation direction, and g_p is the electromagnetic "weighting" function for the p polarization [20, 21].

The relative permittivity of air and water differ considerably for microwave frequencies, therefore even a small amount of air can produce non-trivial changes in the relative permittivity of the resulting mixture. To account for the effect of entrained air, ideally we need to know the fraction of air in water (void fraction) and the vertical distribution of the bubble clouds carrying the air into water. Considering the lack of detailed information about the bubble cloud distribution and the small penetration depth of microwaves, for example, the skin depth is about 2 mm at 10 GHz [22], Hwang [10] uses the fraction of whitecap coverage as a proxy of the void fraction in the mixing rule for the evaluation of the effective relative permittivity ε_e. The approach renders the emission problem of foam-covered ocean surface to two-dimensional (horizontal). The whitecap fraction represents an upper bound of the void fraction because it is equivalent to assuming 100% of air in the depth of microwave influence under the foamy area, whereas the air entrainment decreases exponentially with water depth [23, 24]. The effective relative permittivity ε_e of the air-water mixture is computed with the quadratic mixing rule in a similar approach as that described in Anguelova [25]:

$$\varepsilon_e = \left[f_a \varepsilon_a^{1/2} + (1 - f_a) \varepsilon_{sw}^{1/2} \right]^2, \tag{A5}$$

where ε_{sw} is relative permittivity of sea water without whitecaps and $\varepsilon_a = 1$ is relative permittivity of air, and f_a is air fraction approximated by the whitecap fraction.

Based on this "equivalent medium" approach, the calculated results are in very good agreement with a global dataset of WindSat microwave radiometer measurements with wind speed coverage up to about 42 m/s [26]. The WindSat includes five microwave frequencies (6, 10, 18, 23 and 37 GHz) for both vertical and horizontal polarizations; see figure 3 in [10], figure 9 in [27] and figure 5 in [28]. The agreement between the model and present measurements is also very good (figure A1).

Appendix B. Wind wave energy dissipation

To emphasize the relevance to wave breaking, the combination of wind and wave data are represented by the wind sea energy dissipation rate [11, 13]. The separation of wind sea and swell components is based on the spectrum integration method [12]. The result of applying the method to the FLIP data is shown in figure B1: (a) displays the time-frequency spectral plot; (b) the significant wave heights of the wind sea and swell components; and (c) the peak wave periods of the wind sea and swell components. There is a gap (the dark band in the spectral plot of figure B1a) in the wave data between t = 118.043 and 118.143 d when the pressure sensor depth is adjusted from 3 m to 4 m to prevent breaching. Another adjustment to 5 m is done near t = 119 d without stopping data recording. The time series of wind speed and wind sea peak frequency f_{pw} obtained from swell sea separation are superimposed on the wave spectrum plot (figure B1a) to show the complex evolution of the surface wave spectrum in response to fluctuating wind forcing.

The wind sea component of the wave spectrum is used to estimate the spectrally integrated energy dissipation rate attributed to wave breaking [11]:

$$E_t = \alpha \rho_a U_{10}^3, \text{ with } \alpha = 0.20 \omega_{\#}^{3.3} \eta_{\#} . \quad \text{(B1)}$$

References

[1] Savelyev I B *et al* 2014 *Atmos. Chem. Phys.* **14** 11611
[2] Ovadnevaite J *et al* 2012 *J. Geophys. Res.* **117** D16201.
[3] Callaghan A *et al* 2008 *Geophys. Res. Lett.* **35** L23609.
[4] Liu Y *et al* 1998 *IEEE Trans. Antennas Propag.* **46** 27
[5] Frasier S J *et al* 1998 *J. Geophys. Res.* **103** 18745
[6] Hwang P A and Wang D W 2004 *Geophys. Res. Lett.* **31** L15301
[7] Hwang P A *et al* 2008 *J. Geophys. Res.* **113** C02003
[8] Hwang P A *et al* 2008 *J. Geophys. Res.* **113** C09017
[9] Hwang P A *et al* 1989 *J. Fluid Mech.* **202** 177
[10] Hwang P A 2012 *IEEE Trans. Geos. Rem. Sens.* **50** 2978
[11] Hwang P A and Sletten M A 2008 *J. Geophys. Res.* **113** C02012 (Corrigendum **114** C02015)
[12] Hwang P A *et al* 2012 *J. Atmos. Oceanic Tech.* **29** 116
[13] Hwang P A 2009 *J. Geophys. Res.* **114** C11011
[14] Phillips O M 1984 *J. Phys. Oceanogr.* **14** 1425
[15] Rapp R J and Melville W K 1990 *Phil. Trans. R. Soc. Lond.* **A331** 735
[16] Eastwood C D *et al* 2004 *J. Fluid Mech.* **502** 309
[17] Monahan E C *et al* 1986 *Oceanic Whitecaps and Their Role in Air-sea Exchange Processes* eds Monahan E C and MacNiocaill G (Springer)
[18] Lewis E R and Schwartz S E 2004 *Sea Salt Aerosol Production: Mechanisms, Methods, Measurements, and Models – A Critical Review* (Washington DC: American Geophysical Union)
[19] Smith M H *et al* 1993 *Q. J. R. Meteorol. Soc.* **119** 809
[20] Johnson J T and Zhang M 1999 *IEEE Trans. Geos. Remote Sens.* **37** 2305
[21] Yueh S H *et al* 1994 *Radio Sci.* **29** 799
[22] Plant W J 1990 *Surface Waves and Fluxes, Vol. II – Remote Sensing* eds Geernaert G L and Plant W L (Kluwer Academic)
[23] Wu J 1981 *J. Geophys. Res.* **86** 457

[24] Hwang P A *et al* 1990 *J. Phys. Oceanogr.* **20** 19
[25] Anguelova M D 2008 *J. Geophys. Res.* **113** C08001
[26] Meissner T and Wentz F J 2004 *IEEE Trans. Geosci. Remote Sens.* **42** 1836
[27] Hwang P A *et al* 2013 *J. Atmos. Oceanic Tech.* **30** 2168
[28] Hwang P A and Fois F 2015 *J. Geophys. Res.* **120** 3640

The relationship between ocean surface turbulence and air-sea gas transfer velocity: An in-situ evaluation

L Esters[1], S Landwehr[1], G Sutherland[2], T G Bell[3], E S Saltzman[4], K H Christensen[5], S D Miller[6] and B Ward[1]

[1] AirSea Laboratory, NUI, Galway, Ireland
[2] Department of Mathematics, University of Oslo, Norway
[3] Plymouth Marine Laboratory, United Kingdom
[4] Department of Earth Sciences, University of California, Irvine, CA, USA
[5] Norwegian Meteorological Institute, Oslo, Norway
[6] Atmospheric Sciences Research Center, State University of New York at Albany, USA

E-mail: bward@nuigalway.ie

Abstract.
Although the air-sea gas transfer velocity k is usually parameterized with wind speed, the so-called small-eddy model suggests a relationship between k and ocean surface dissipation of turbulent kinetic energy ϵ. Laboratory and field measurements of k and ϵ have shown that this model holds in various ecosystems. Here, field observations are presented supporting the theoretical model in the open ocean. These observations are based on measurements from the Air-Sea Interaction Profiler and eddy covariance CO_2 and DMS air-sea flux data collected during the Knorr11 cruise. We show that the model results can be improved when applying a variable Schmidt number exponent compared to a commonly used constant value of 1/2. Scaling ϵ to the viscous sublayer allows us to investigate the model at different depths and to expand its applicability for more extensive data sets.

1. Introduction

Gas fluxes across the air-sea interface are computed according to:

$$F = K\ s\ \Delta P \tag{1}$$

where ΔP is the partial pressure difference of the particular gas between the ocean and the atmosphere, s is the solubility of the gas, and K is the total gas transfer velocity [1, 2]. The technology for field measurements of ΔP, of both DMS and CO_2, has matured to the point where they are readily accessible, and their values can be determined with high precision. Therefore, the main challenge remaining is to estimate K to accurately model air-sea fluxes, as K is one of the dominant sources of uncertainty in the calculation of these fluxes (e.g., [3]).

The water side air-sea gas transfer velocity k is usually parameterized as a function of wind speed ($k = f(u10)$) [4], as this is a readily available parameter from in-situ and satellite observations. However, wind speed is not the only factor influencing k. As the physical processes that influence k are mostly controlled by the near-surface dissipation rate of turbulent kinetic

Content from this work may be used under the terms of the Creative Commons Attribution 3.0 licence. Any further distribution of this work must maintain attribution to the author(s) and the title of the work, journal citation and DOI.
Published under licence by IOP Publishing Ltd

IOP Conf. Series: Earth and Environmental Science **35** (2016) 012005 doi:10.1088/1755-1315/35/1/012005

energy ϵ [5], attempts have been made to directly relate k to ϵ, originally described by surface renewal models [6, 7]. These models describe the air-sea gas exchange by linking the diffusive exchange at the interface with the continuous disruption of the diffusive boundary layer due to turbulent motions. Higher turbulence means more frequent renewal of the water side boundary layer and thus a higher gas exchange rate. Lamont and Scott [8] suggest to model this relation using the small-eddy model:

$$k = A\,\mathrm{Sc}^{-n}(\epsilon_0\ \nu)^{1/4} \tag{2}$$

where A is a proportionality constant, ϵ_0 the dissipation rate of TKE at the air-sea interface, Sc is the Schmidt number, which is defined as the ratio of the kinematic viscosity of water ν to mass diffusivity of water D so, $\mathrm{Sc} = \nu/D$, and n is the Schmidt number exponent. For the surface renewal models, Lamont and Scott [8] predicted a Schmidt number exponent of $n = 1/2$. This Schmidt number exponent of $n = 1/2$ is not applicable for interfaces covered with surfactants. For film-covered surfaces, a Schmidt number exponent of $n = 2/3$ is found [9]. In reality, sea surfaces might be partially covered with surfactants and n varies between 1/2 for a wavy surface and 2/3 for a flat surface depending on the sea surface conditions [10, 11]. Laboratory experiments [12, 13] and field measurements [14, 15, 16, 17] of k and ϵ have shown that this theoretical relation agrees with observations, and that ϵ is a good predictor of gas transfer in various ecosystems. These studies cover various types of environmental systems and range from tank experiments to lakes, rivers, estuaries, and coastal oceans as well as the model-world Biosphere 2 [14, 15, 16, 17]. In addition, these studies cover various types of environmental forcing like wind, tides, and rain and were measured using a variety of instrumentation. This variety of observational setups and environments demonstrates the universality of the small-eddy model described in equation 2.

All these studies show a high correlation between k and ϵ, though significant variation is found in the empirical proportionality coefficient A (equation 2), usually calculated through linear regression. As the turbulence measurements were carried out at various depths below the surface, the differences in A are most likely related to this, as ϵ strongly depends on depth in the surface region of the ocean [18]. Another source of variability in A is the sea surface condition as well as surfactants at the sea surface (e.g., [12]), but Tokoro *et al.* [15] argue that this should affect ϵ and not A. Wang *et al.* [17] evaluated the small-eddy model under non-breaking wave conditions based on turbulence measurements directly below the sea surface. Their results indicate that A scales linearly with $\log(\epsilon)$ and they show that the relation holds for other published studies focusing on similar environments (i.e., lakes with wind and wave-induced surface turbulences), and in laboratory experiments. However, this relation contradicts the measurements of Zappa *et al.* [14], which were taken in different environmental systems with different forcing.

To the best of our knowledge, no investigations have been performed in the open ocean relating k to ϵ. Measurements of eddy covariance CO_2 and DMS air-sea fluxes and of ϵ from the Air-Sea Interaction Profiler (ASIP), taken during the Knorr11 cruise, provide an opportunity to investigate the relation between k and ϵ in the open waters of the North Atlantic Ocean.

2. Methods

Measurements were taken during a field campaign in the North Atlantic, aboard the R/V *Knorr* from late June to mid-July 2011 leaving and returning to Woods Hole, USA [19, 20, 21, 22, 23].

2.1. Flux measurements

The 3D wind speed was measured at 10 Hz with two Csat3 sonic anemometers mounted at the bow mast of the R/V *Knorr*. The measurements were corrected for ship motion as described in Miller *et al.* [24] and Landwehr *et al.* [25]. The time series of the motion corrected wind speeds were used to calculate the air-side friction velocity u_{*_a} and the trace gas fluxes. The CO_2 fluctuations were measured at 10 Hz with a non-dispersive infra-red gas analyzer of the brand LICOR. A diffusion dryer was used to remove ambient water vapor fluctuations [26]. It was shown in Landwehr *et al.* [27] that this significantly improves the quality of the direct CO_2 flux measurements. The DMS flux measurements and data analysis are described in Bell *et al.* [28]. The total gas transfer velocities K of CO_2 and DMS were calculated using equation 1 based on 10-min averaged measured data. These K-values combine the effects of processes at the air as well as the water side of the air-sea interface. The water side gas transfer velocity k was calculated and normalized to a Schmidt number of 660 (CO_2 at 25°C):

$$k_{660} = k \cdot \left(\frac{660}{\mathrm{Sc}_x}\right)^{-0.5} \tag{3}$$

where Sc_x refers to the Schmidt number at the in situ seawater state for either DMS or CO_2.

2.2. ASIP

Presented here are data from four deployments (a total of 283 profiles) of ASIP, an autonomous upwardly rising microstructure profiler. ASIP, which is extensively described in Ward *et al.* [29], is equipped with a variety of sensors and ϵ is calculated from two shear probes. To validate the small-eddy model only the surface values of ϵ are of interest. As the vertical resolution of the measured ϵ is approximately 0.5 m, all data points within the uppermost 0.5 m of the ocean are declared as surface values ($\epsilon_{0.5}$). These surface values of ϵ are compared to the measured k_{660} based on the DMS and CO_2 flux measurements (in the following called k_{DMS} and k_{CO_2}).

2.3. Scaling ϵ

It is not straightforward to measure ϵ at the air-sea interface and there is little open-ocean data of near-surface ϵ available. Therefore, it is helpful to scale ϵ close to the air-sea interface, once a universal relation between ϵ and k is determined. Lorke and Peeters [5] took the approach to consider the oceanic boundary layer as a shear-driven, flat surface, and scaled ϵ with the *law of the wall* (LOW) [30]. The underlying assumption of this approach is that the total stress in the boundary layer is constant, which leads to a log-linear velocity profile:

$$\epsilon(z) = \frac{u_{*_w}^3}{\kappa z} \tag{4}$$

where $\kappa \approx 0.41$ is the von Kármán constant, z is the distance down from the ocean surface, and u_{*_w} is the measured water side friction velocity. The applicability of the assumption that the sea surface is purely shear driven is questioned [31, 32, 33, 34]. Measurements show that higher dissipation rates can be found than those predicted by the LOW. These higher rates are usually ascribed to breaking waves [31, 32, 33]. The results of this study show that the scaled ϵ are nevertheless well correlated to the measured ϵ from ASIP. By comparing the scaled and measured surface ϵ an offset between them was determined. The profiles of ϵ were extrapolated to the viscous sublayer thickness, which has been shown to be a function of wind speed [35, 36]. The relationship from Wu [37] was used to determine this thickness, given by $z_\nu = 11\nu/u_{*_w}$, and an offset was subtracted to yield the dissipation rate at the interface (ϵ_0). These interface values are on average two orders of magnitude higher than the actual measured $\epsilon_{0.5}$ in the uppermost 0.5 m of the ocean.

2.4. Schmidt number exponent

All mentioned field studies on the small-eddy model assumed a constant Schmidt number exponent of $n = 1/2$ for the full range of their measurements [14, 15, 16]. However, n depends on the sea surface conditions and is known to vary between 1/2 for a wavy sea surface to 2/3 for a smooth surface [10, 11]. Krall [38] carried out experiments investigating the dependence of n on wind speed for different surfactant coverage in the Aeolotron at the university in Heidelberg. During these experiments, Krall [38] distinguished between three surface conditions: clean surface, 0.052 mmol/l Triton and 0.26 mmol/l Triton, and determined different $n(u_{10})$ relations for each of them (figure 1). The higher the surfactants coverage, the higher is the wind speed at which n reaches 2/3, and the steeper is the wind speed dependence.

Neither the surfactant coverage nor the mean square slope were measured during Knorr11, so n was parameterized based on wind speed at 10 m height (u_{10}) only. It is assumed that a medium surfactant coverage prevailed during Knorr11, as there were several instances of high chlorophyll values.

The best fit of $n(u10)$ was determined for k_{DMS} and for k_{CO_2} based on equation 2 with the assumption that the proportionality would be unity ($A = 1$). The best fit for $n(u_{10})$ based on k_{DMS} is $n = -0.259\log(u_{10}) + 0.83$ and $n = -0.162\log(u_{10}) + 0.642$ based on k_{CO_2}. For both gases a logarithmic fit best described the relationship over a linear fit. Overall, the relation based on k_{DMS} yielded better results (RMSD ± 4.3 cm hr^{-1}) than the one based on k_{CO_2} (RMSD ± 21.5 cm hr^{-1}) due to less variability in k_{DMS} measurements. According to figure 1, $n(u10)$ reaches the upper limit at a similar wind speed to the one for medium surfactants [38] and follows its trend whilst being less steep.

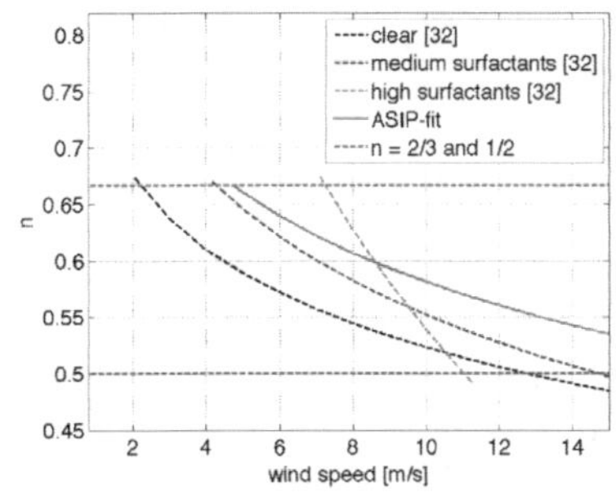

Figure 1. The Schmidt number exponent n versus wind speed based on Krall [38] for high surfactant coverage (0.26 mmol/l Triton in green), medium surfactant coverage (0.052 mmol/l Triton in blue), clean surface (black), and the best fit between the measured ϵ and k_{DMS} in the small-eddy model during Knorr11 (red). The grey colored lines give the theoretical boundaries of $n = 1/2$ and $n = 2/3$.

3. Results and discussion

To improve the statistics, the ASIP data are averaged over 90 minutes. Figure 2a and 2b show the results of applying $n = f(u10)$ instead of a constant $n = 1/2$ value in equation 2. For $n = f(u10)$, 89% of the variation in k_{DMS} can be explained by the small-eddy model, compared to 53% for a constant n. The k_{CO_2} data are more variable than k_{DMS} (figure 2c and 2d). For $n = 1/2$, R^2 is found to be 0.16, whereas for $n = f(u10)$, $R^2 = 0.2$. For both gases, DMS and CO_2, k is highly variable at low wind speeds, and thus higher n.

The proportionality coefficient A, which is found through data regression, has to be adapted when applying $n(u10)$ instead of a constant $n = 1/2$. Figure 2b shows that $A = 1.0 \pm 0.11$ for k_{DMS}, which is reasonable as the parametrization of $n(u10)$ is based on the assumption that $A = 1$ in equation 2. For the commonly used $n = 1/2$ a proportionality of $A = 0.42 \pm 0.05$ is observed for k_{DMS}. This value is nearly identical to the theoretical value of 0.4 [8] and close to values found in field experiments of Zappa *et al.* [14] ($A = 0.419 \pm 0.130$), whose depth measurements for ϵ ranged from a few cm to 3 m below the sea surface. Vachon *et al.* [16] estimated $A = 0.44 \pm 0.01$ for large lakes and $A = 0.39 \pm 0.02$ for smaller lakes measuring ϵ at 10 cm depth by using the floating chamber method. Thus, when applying the same assumptions as in these studies in lakes, rivers, and estuaries, similar results are obtained in the open ocean.

The proportionality constant A for $n = 1/2$ (k_{DMS}) is, however, approximately double that of Tokoro *et al.* [15], who determined values of $A = 0.13$ to 0.2 at depths scaled to 86 cm and

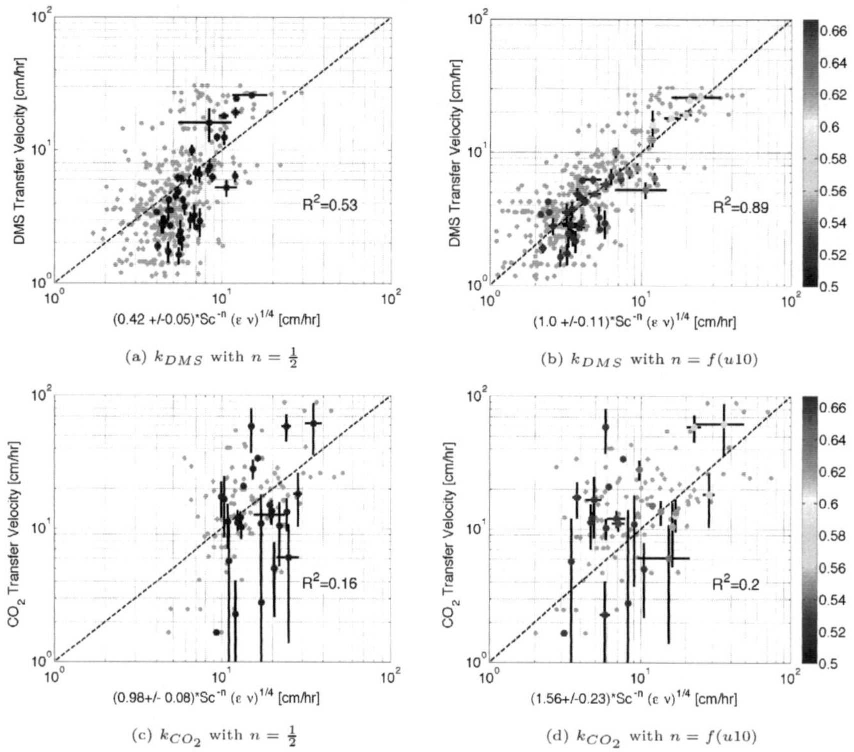

Figure 2. Ninety-minute averaged k and the prediction of the small-eddy model based on measured ϵ taken within the uppermost 0.5 m ($\epsilon_{0.5}$) of the ocean in log–log space. The color code represents n, the errors bars show the standard error of k (std/sample size$^{1/2}$), the grey data in the background are the non-averaged data, and the dashed line represents the 1:1 relation.

also double that of Wang *et al.* [17], who report values of $A = 0.25$ measured at 0.25 cm depth and $A = 0.31$ at 10 cm depth. To further investigate this variability in A, the measured ϵ are extrapolated closer to the air-sea interface (the viscous sublayer). These scaled and extrapolated values of ϵ feature 6% less variability than those originally measured. This reduction in variability might explain the higher coefficient of determination R^2 at the interface compared to the total uppermost 0.5 m (figure 2 compared to figure 3). For $n = f(u10)$, the small-eddy model explains 91% of the variability in k_{DMS} at the interface (ϵ_0) and 69% for $n = 1/2$. Again, the relation between k_{CO_2} and the right-hand side of the small-eddy model is lower than for k_{DMS}, and the small-eddy model explains only 18% of the variability for $n = 1/2$ and 19% for $n = f(u10)$.

For k_{DMS} as well as for k_{CO_2}, the determined A have lower values when scaled to the air-sea interface compared to the measurements at 0.5 m (0.18 ± 0.03 for $n = 1/2$ and 0.39 ± 0.06 for $n(u10)$ based on k_{DMS}, and 0.31 ± 0.09 for $n = 1/2$ and 0.47 ± 0.15 based on k_{CO_2}). The value of A for k_{DMS} with $n = 1/2$ at the interface is significantly closer to those reported by Wang *et al.* [17] measured at 0.25 cm depth than the A found for the same setting within the whole surface layer. This supports the idea that A depends on the depth of the turbulence measurements, where A has to follow the $\epsilon(z)$ dependency to balance the small-eddy model.

Once a parametrization for $n(u10)$ and a corresponding A are determined, the small-eddy model is applicable to more extensive gas transfer datasets. This approach can be used to predict k by figure of ϵ being scaled based on LOW and thus based on u_{*w} instead of real ϵ measurements. Figure 4 shows the prediction of the small-eddy model for the complete Knorr11

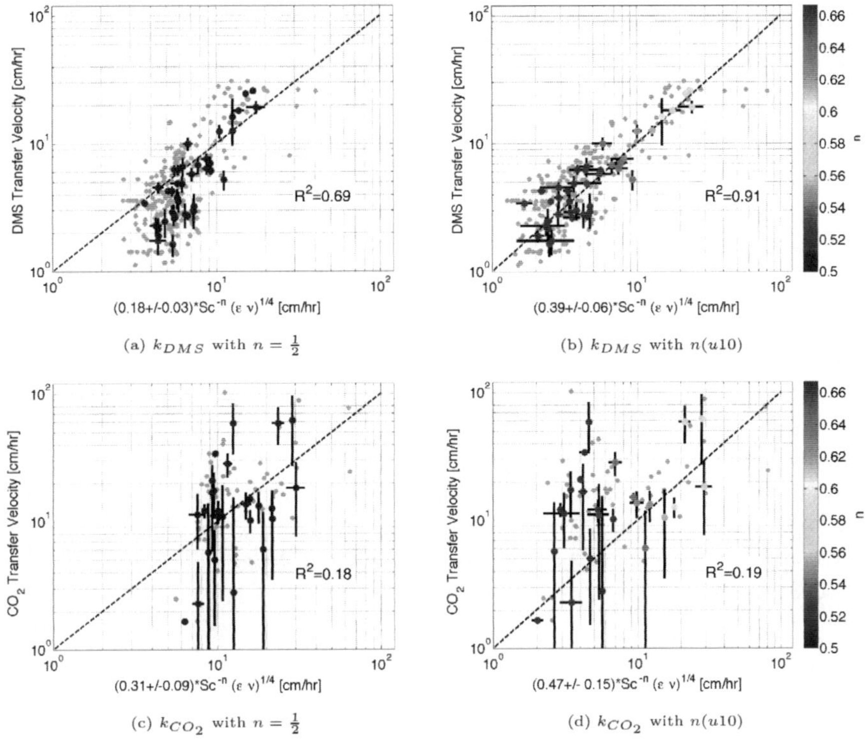

Figure 3. Ninety-minute averaged k and the prediction of the small-eddy model based on $\epsilon(z)$ predicted by the LOW (equation 4), fitted to the measured ϵ profiles and extrapolated to the viscous sublayer depth ϵ_0 ($z_\nu = 11\nu/u_{*_w}$). The colour code represents n, the errors bars show the standard error of k (std/sample size$^{1/2}$), the grey data in the background are the non-averaged data, and the dashed line represents the 1:1 relation.

data [20].

Bin averaging k reduces variability and results in a linear relationship between k_{CO_2} and the right hand side of equation 2, with a R^2 value of 0.92. For k_{DMS} a lower correlation is found, as there is a deviation in the transfer velocity at high wind speeds, which has been explained by Bell *et al.* [20]. If these data points are ignored ($u10 > 12$ m/s), R^2 reaches 0.78, so the small-eddy model explains a large part of the variability in k_{DMS} for low and medium wind speeds.

No measurements of ϵ exist at high wind speeds during Knorr11 (indicated in figure 4). Therefore, a reduction in ϵ during high wind speeds could explain the discrepancy between the predicted k_{DMS} and the measured one (figure 4). A reduction in ϵ with a simultaneous increase in bubbles could also explain the different behavior in k_{DMS} and k_{CO_2} [20].

4. Conclusions

Various studies on gas exchange across the air-sea interface, including a variety of methods and techniques ranging from theoretical approaches to laboratory studies to field campaigns in lakes, coastal areas and estuaries, have verified ϵ to be a good predictor for k using the small-eddy model [5, 12, 13, 14, 15, 16]. Our measurements of k and ϵ verify that this theoretical relation holds well in the open ocean.

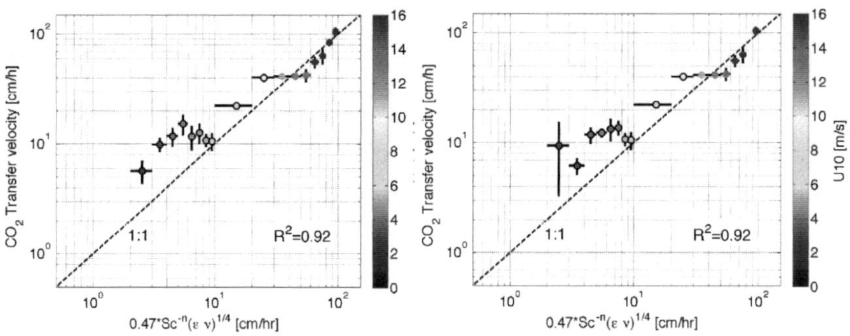

(a) k_{DMS} (b) k_{CO_2}

Figure 4. Bin-averaged measured k against the u_{*_w}-based prediction (LOW in equation 4) of the small-eddy model for the full Knorr11 data. Values of ϵ_0 are predicted by the LOW for the viscous sublayer depth ($z_\nu = 11\nu/u_*$) with a determined offset between measurements and predictions being subtracted. For the small-eddy model a varying Schmidt number exponent $n = f(u10)$ is applied. Only the circled data points indicate bins in which more than three actual turbulence measurements exist. For wind speeds higher than 12 m/s, we do not have any direct measurements of ϵ. The colour code represents the wind speed, the errors are given by the standard error of k (std/bincounts$^{1/2}$) and the dashed line represents the 1:1 relation.

When applying the same assumption that $n = 1/2$, as in former studies, the small-eddy model explains 53% of the variability in k_{DMS} and only 16% of the variability in k_{CO_2}. Our field data show that this relation can be significantly improved when applying a varying Schmidt number exponent n. This n is know to vary with the sea surface conditions. A parametrization for n based on wind speed improves the predictability of the small-eddy model to 89% for k_{DMS}. This $n = f(u10)$ changes the right-hand side of equation 2 similar to the approach of Wang *et al.* [17], who scale the proportionality coefficient with $A \propto log(\epsilon)$. In both cases, the expression is shifted towards lower values for low ϵ and vice versa.

Wang *et al.* [17] report different logarithmic parameterizations of A for different depths to retain a linear relation for the small-eddy model. The parametrization of $n = f(u10)$ presented here, however, is universal for different depth levels. In this case, the constant A has to only be adapted for each depth level to avoid any offset in the linear relation between both sides of the small-eddy model. Thus our results support the idea that A depends on depth, and increases towards the air-sea interface as A has to balance the increase in ϵ towards the interface. Measurements of ϵ reported in earlier studies on this topic were taken at various depths, which makes comparisons of A difficult.

Surface conditions and the function for n (either a constant or $n = f(u_{10})$) impact values of A. Once an appropriate function for n and a corresponding A for the depth of interest is found, a universal relation between k and ϵ can be determined.

In this work the LOW is used to scale ϵ from depth measurements to the air-sea interface, but future work will involve the implementation of different scaling approaches to predict ϵ [21].

Acknowledgments

Leonie Esters gratefully receives support from the NUIG College of Science Fellowship Programme. The attendance of the GTWS-7 was supported by the Marine Institute Networking and Travel Grant, the Ryan Institute Travel Support Scheme, and the GTWS-7 travel support scheme. Funding for the Knorr11 field experiment was supported by Science Foundation Ireland under grant number 08/US/I1455 and the EU FP7 project CARBOCHANGE under grant agreement no. 264879.

References

[1] Weiss R F 1974 Carbon dioxide in water and seawater: The solubility of a nonideal gas *Mar. Chem.* **2** 203–215

[2] Wanninkhof R, Asher W E, Ho D T, Sweeney C and McGillis W R 2009 Advances in Quantifying Air–Sea Gas Exchange and Environmental Forcing *Ann. Rev. Mar. Sci.* **1** 213–244

[3] Johnson M T, Hughes C, Bell T G and Liss P S 2011 A Rumsfeldian analysis of uncertainty in air-sea gas exchange *Gas Transfer at Water Surfaces 2010* ed Komori S, McGillis W and Kurose R (Kyoto, Japan: Kyoto University Press.) pp 464–484

[4] Wanninkhof R 1992 Relationship between wind speed and gas exchange over the ocean *J. Geophys. Res.* **97** 7373–7382

[5] Lorke A and Peeters F 2006 Toward a Unified Scaling Relation for Interfacial Fluxes *J. Phys. Oceanogr.* **36** 955–961 10.1175/JPO2903.1

[6] Higbie R 1935 The rate of absorption of a pure gas into a still liquid during short periods of exposure *AlChE J.* **31** 365–389

[7] Danckwerts P V 1951 Significance of liquid film coefficients in gas absorption *Ind. Eng. Chem.* **43** 1460–1467

[8] Lamont J C and Scott D S 1970 An eddy cell model of mass transfer into the surface of a turbulent liquid *AlChE J.* **16** 513–519

[9] Davies J T 1972 *Turbulence phenomena* (Academic Press., N.Y.)

[10] Deacon E L 1977 Gas transfer to and across an air–water interface *Tellus* **29** 363–374

[11] Jähne B, Henz G and Dietrich W 1987 Measurement of the diffusion coefficients of sparingly soluble gases in water *J. Geophys. Res.* **92** 10767–10776

[12] Asher W and Pankow J F 1986 The interaction of mechanically generated turbulence and interfacial films with a liquid phase controlled gas/liquid transport process *Tellus B* **38** 305–318

[13] Moog D B and Jirka G H 1999 Air-Water Gas Transfer In Uniform Channel Flow *J. Hydraul. Eng.* **125** 3–10

[14] Zappa C J, McGillis W R, Raymond P A, Edson J B, Hintsa E J, Zemmelink H J, Dacey J W H and Ho D T 2007 Environmental turbulent mixing controls on air–water gas exchange in marine and aquatic systems *Geophys. Res. Lett.* **34** L10601

[15] Tokoro T, Kayanne H, Watanabe A, Nadaoka K, Tamura H, Nozaki K, Kato K and Negishi A 2008 High gas–transfer velocity in coastal regions with high energy-dissipation rates *J. Geophys. Res.* **113** C11006

[16] Vachon D, Prairie Y T and Cole J J 2010 The relationship between near–surface turbulence and gas transfer velocity in freshwater systems and its implications for floating chamber measurements of gas exchange *Limnol. Oceanogr.* **55** 1723–1732

[17] Wang B, Liao Q, Fillingham J H and Bootsma H A 2015 On the coefficients of small eddy and surface divergence models for the air–water gas transfer velocity *J. Geophys. Res. Oceans* **120** 2129–2146

[18] Sutherland G, Ward B and Chrisensen K H 2013 Wave–turbulence scaling in the ocean mixed layer *Ocean Sci.* **9** 597–608

[19] Scanlon B and Ward B 2013 Oceanic wave breaking coverage seperation techniques for active and maturing whitecaps *Methods Oceanogr.* **8** 1–12

[20] Bell T G, De Bruyn W, Miller S D, Ward B, Christensen K H and Saltzman E S 2013 Air–sea dimethylsulfide (DMS) gas transfer in the North Atlantic: evidence for limited interfacial gas exchange at high wind speed *Atmos. Chem. Phys.* **13** 11073–11087

[21] Sutherland G, Christensen K H and Ward B 2013 Wave-turbulence scaling in the ocean mixed layer *Ocean Sci.* **9** 597–608

[22] Christensen K H, Röhr J, Ward B, Fer I, Broström G, Sætra Ø and Breivik Ø 2013 Surface wave measurements using ship mounted ultrasonic altimeter *Method Oceanogr.* **6** 1–15

[23] Sutherland G, Christensen K H and Ward B 2014 Evaluating langmuir turbulence in the oceanic boundary layer *J. Geophys. Res.* **119** 1899–1910

[24] Miller S, Marandino C, van de Bruyn W and Saltzman E S 2009 Air-sea gas exchange of co_2 and dms in the north atlantic by eddy covariance *Geophys. Res. Lett.* **36** L15816

[25] Landwehr S, O'Sullivan N and Ward B 2015 Direct Flux Measurements from Mobile Platforms at Sea: Motion and Airflow Distortion Corrections Revisited *J. Atmos. Oceanic Technol.* **32** 1163 – 1178

[26] Miller S D, Marandino C and Saltzman E S 2010 Ship-based measurement of air-sea CO_2 exchange by eddy covariance *J. Geophys. Res.* **115** D02304

[27] Landwehr S, Miller S D, Smith M J, Saltzman E S and Ward B 2014 Analysis of the pkt correction for direct co_2 flux measurements over the ocean *Atmospheric Chemistry and Physics* **14** 3361–3372

[28] Bell T G, Bruyn W D, Miller S D, Ward B, Christensen K and Saltzman E S 2013 Air/sea DMS gas transfer

in the North Atlantic: evidence for limited interfacial gas exchange at high wind speed *Atmos. Chem. Phys.* 11073–11087

[29] Ward B, Fristedt T, Callaghan A H, Sutherland G, Sanchez X, Vialand J and tenDoeschate A 2014 The air–sea interaction profiler for microstructure measurements in the upper ocean *J. Atmos. Oceanic Technol.* **31** 2246–2267

[30] Callaghan A H, Ward B and Vialard J 2014 Influence of surface forcing on near-surface and mixing layer turbulence in the tropical indian ocean *Deep-Sea Res.* **4**

[31] Kitaigorodskii S A, Donelan M A, Lumley J L and Terray E A 1983 Wave–Turbulence Interactions in the Upper Ocean. Part II. Statistical Characteristics of Wave and Turbulent Components of the Random Velocity Field in the Marine Surface Layer *J. Phys. Oceanogr.* **13** 1988–1999

[32] Agrawal Y C, Terray E A, Donelan M A, Hwang P A and Williams III A J 1992 Enhanced dissipation of kinetic energy beneath surface waves *Nature* **359** 219–220

[33] Terray E A, Donelan M A, Agrawal Y C, Drennan W M, Kahma K K, Williams III A J, Hwang P A and Kitaigorodskii S A 1996 Estimates of Kinetic Energy Dissipation under Breaking Waves *J. Phys. Oceanogr.* **26** 792–807

[34] Babanin A V and Haus B K 2009 On the Existence of Water Turbulence Induced by Nonbreaking Surface Waves *J. Phys. Oceanogr.* **39** 2675–2679

[35] Ward B and Donelan M A 2006 Thermometric measurements of the molecular sublayer at the air–water interface *Geophys. Res. Lett.* **33** L07605

[36] Ward B 2007 *Transport at the Air Sea Interface — Measurements, Models and Parameterizations* ed Garbe C S, Handler R A and Jähne B (Springer Verlag)

[37] Wu J 1971 An estimation of oceanic thermal–sublayer thickness *J. Phys. Oceanogr.* **1** 284–286

[38] Krall K E 2013 *Laboratory Investigations of Air-Sea Gas Transfer under a Wide Range of Water Surface Conditions* (PhD. Ruperto-Carola University of Heidelberg, Germany)

DNS and measurements of scalar transfer across an air-water interface during inception and growth of Langmuir circulation

A Hafsi[1], Y Ma[2], M Buckley[2], A E Tejada-Martinez[1] and F Veron[2]

[1]Civil and Environmental Engineering Department, University of South Florida, 4202 E. Fowler Avenue, Tampa, FL 33620-535, USA
[2]School of Marine Science and Policy, College of Earth, Ocean and Environment, University of Delaware, Newark, DE 19716, USA

Email: aetejada@usf.edu and fveron@udel.edu

Abstract. Direct numerical simulations (DNS) of an initially quiescent coupled air-water interface driven by an air flow with free stream speed of 5 m/s have been conducted and scalar transfer from the air side to the water side and subsequent vertical transport in the water column have been analysed. Two simulations are compared: one with a freely deforming interface, giving rise to gravity-capillary waves and aqueous Langmuir turbulence (LT) characterized by small-scale (centimeter-scale) Langmuir cells (LC), and the other with the interface intentionally held flat, i.e., without LC. It is concluded that LT serves to enhance vertical transport of the scalar in the water side and in the process increases scalar transfer efficiency from the air side to the water side relative to the shear-dominated turbulence in the flat interface case. Furthermore, transition to LT was observed to be accompanied by a spike in scalar flux characterized by an order of magnitude increase. These episodic flux increases, if linked to gusts and overall unsteadiness in the wind field, are expected to be an important contributor in determining the long-term average of the air-sea gas fluxes.

1. Introduction

Turbulence associated with small scale (centimeter-scale) eddies known as Langmuir cells (or LCs) (see sketch in figure 1) can have direct impact on gas uptake by oceans. During low wind speeds (1–2 m/s), wave-current interaction leads to the generation of Langmuir turbulence characterized by these small scale LCs. As seen in the sketch of figure 1 (left panel), LC consists of parallel counter-rotating vortices roughly aligned in the direction of the wind. It is well known that LCs result from the interaction between the Stokes drift velocity induced by surface waves and the wind-driven shear current. The panel on the right in figure 1 shows a surface temperature image (with red indicating warmer temperatures and blue colder temperatures) of a wind-driven air-water interface during the development of small-scale Langmuir cells in laboratory experiments [4]. The presence of the Langmuir cells is manifested through the pattern in surface temperature. For example, the blue temperature streaks coincide with the surface convergence (downwelling) zones of the Langmuir cells.

Content from this work may be used under the terms of the Creative Commons Attribution 3.0 licence. Any further distribution of this work must maintain attribution to the author(s) and the title of the work, journal citation and DOI.
Published under licence by IOP Publishing Ltd

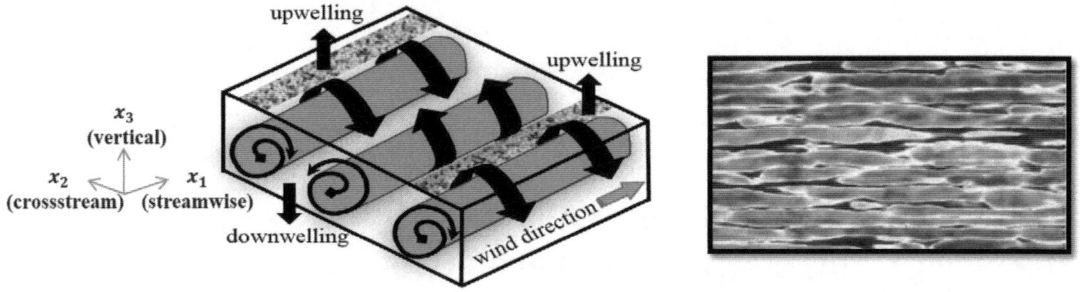

Figure 1. Sketch of LCs on left and surface manifestation of LCs on the air-water interface in terms of temperature on right (Veron and Melville [4]).

In the measurements of Veron and Melville [4] the inception of small-scale LCs and subsequent transition to Langmuir turbulence (prior to the onset of micro-breaking wind waves) lead to a 70% increase in gas transfer velocity (a measure of gas transfer efficiency), demonstrating the strong impact of LC on gas transfer across the air-water interface. Transition to Langmuir turbulence refers to when the initially coherent vortices become unstable and lose coherency as they start to interact with each other nonlinearly. The measurements of Veron and Melville [4] were collected under a gradually increasing wind with final speed of 5 m/s at a fetch of 10.72 m and during the presence of gravity-capillary waves possessing wavelengths up to 10–15 cm. More recent experiments made with sub-surface particle image velocimetry (PIV) and surface infrared radiometry were conducted in the large air-sea interaction facility at the University of Delaware. Several wind forcings were studied in order to obtain a variety of LC scales and intensities. It was found that the LCs provide intense surface renewal (see figure 2) and disrupt the near surface molecular layers. In particular, the LC rapidly transport surface layers to depth thereby transporting momentum away from the surface at rates much larger than that of molecular diffusion. The surface infrared measurements also provide air-water heat transfer velocity estimates. As a result strong correlations between scalar flux estimates and various kinematic quantities obtained from the PIV measurements and derived from moments of the sub-surface velocity gradient tensor were found. Figure 2 presents snapshots from a video filmed in these laboratory measurements taken at two different times, showing the progress of fluorescent dye initially deposited on the surface of a wind-driven air-water interface as a tracer during the presence of small-scale LCs. The images are in the crosswind-vertical plane, with the wind and waves traveling in the "into the page" direction. In figure 2 we can see the formation of pairs of vorticies that correspond to small-scale Langmuir cells. By comparing the progress of the fluorescent dye within the water side between an early time (figure 2a) and a later time (figure 2b) the effect of Langmuir cells on vertical transport of momentum and scalars can be observed.

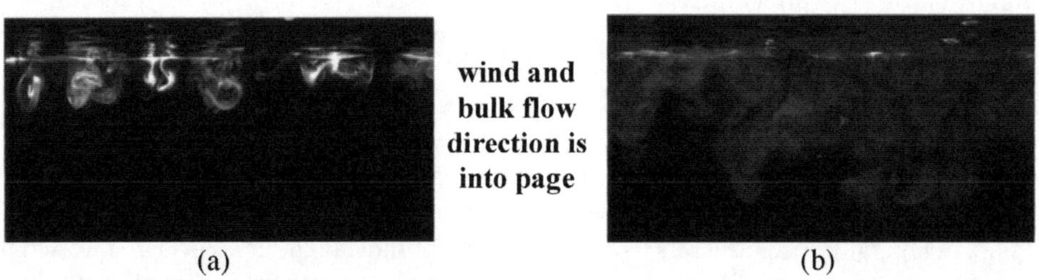

Figure 2. (a) Fluorescent dye sprayed on a wind-driven air-water interface as a tracer during the presence of small-scale Langmuir cells at (a) an early time, (b) later time.

IOP Conf. Series: Earth and Environmental Science **35** (2016) 012006 doi:10.1088/1755-1315/35/1/012006

Komori *et al.* [3] performed a direct numerical simulation (DNS) similar to the laboratory setting of Veron and Melville [4]. In their DNS, LC-like structures coexisting with micro-breaking capillary waves riding on larger gravity-capillary waves were identified. However, a study of LC-like structures and their connection with gas transfer efficiency across the air-water interface was not explored. For the present study, a similar DNS of a wind-driven coupled air-water interface was conducted focusing on the vertical transport induced by small scale LCs to further establish the connection between the cells and scalar transfer across the interface.

2. DNS equations

The DNS equations consist of the incompressible continuity and the Navier–Stokes equations

$$\frac{\partial u_i}{\partial x_i} = 0 \tag{1}$$

$$\frac{\partial u_i}{\partial t} + u_j \frac{\partial u_i}{\partial x_j} = -\frac{1}{\rho}\frac{\partial p}{\partial x_i} + \nu \frac{\partial^2 u_i}{\partial x_j^2} \tag{2}$$

where u_i denotes the velocity field [i = 1, 2 and 3 represent the streamwise (or downwind), spanwise and vertical components, respectively], p is the pressure field, ν is the molecular kinematic viscosity and ρ is the density. In addition to the continuity and momentum (Navier–Stokes) equations to predict air and water flow velocities and pressures, the concentration of dissolved scalar (i.e., gas) in air and in water is predicted via an additional scalar transport equation:

$$\frac{\partial C}{\partial t} + u_j \frac{\partial C}{\partial x_j} = \kappa \frac{\partial^2 C}{\partial x_j^2} \tag{3}$$

where C is the concentration of the scalar and x is the molecular diffusivity of the scalar. In this study, as has been done in others, the ratio ν/κ or Schmidt number is set to 1 because this does not affect the fundamental turbulent mechanisms promoting scalar transfer across the interface.

The flow solver used in this research is based on the open source numerical collection of C++ libraries OpenFOAM (Open source Field Operations and Manipulations), which uses the finite volume method to discretize the continuity, Navier–Stokes and passive scalar equations. After the discretization of the Navier–Stokes equations, the resulting pressure-velocity coupling for the momentum equation is resolved using the pressure implicit with splitting of operators (PISO) method [2]. The diffusion terms are discretized using a central finite difference method whereas the nonlinear advection terms are discretized using a second order accurate upwind method. InterFOAM, a solver included with OpenFOAM, is used in order to track the air-water interface via a modified volume of fluid multiphase methodology.

3. Results and discussion

Simulations conducted follow closely the recent DNS of Komori *et al.* [3] of wind-driven, coupled air-water molecular boundary layers with a nonlinearly deformable surface (interface). The juxtaposed domains of the air and water fluids along an initially flat gas-liquid interface are represented by rectangular boxes with depths of δ and 2δ, respectively, where δ = 1.25 cm, while the streamwise and spanwise lengths are 8δ and 4δ, respectively (see figure 3a). A Cartesian coordinate system was adopted where the streamwise and spanwise directions are given by the x-axis and the y-axis, respectively, while the z-axis spans the air and water depths (see computational domain in figure 3a).

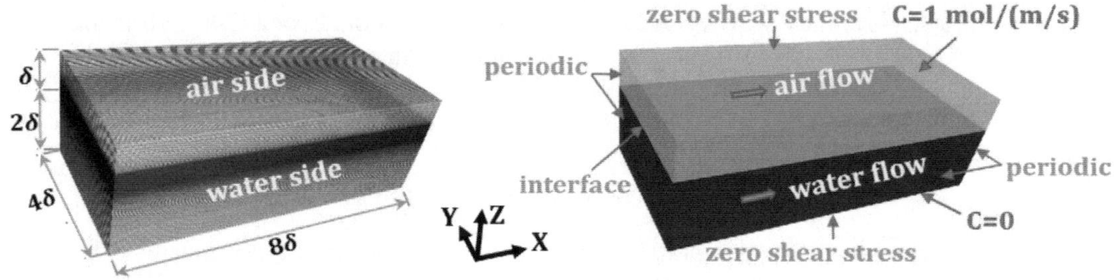

Figure 3. (a) Coupled air-water flow domain and computational mesh and (b) flow domain and boundary conditions: zero shear stress with zero normal flow and prescribed scalar concentration at top and bottom of the domain; periodicity in streamwise and spanwise directions.

The computational mesh shown in figure 3a consisted of 200 by 100 by 60 grid points on the air side and 200 by 100 by 120 grid points on the water side in the streamwise, spanwise and vertical directions, respectively. The computational mesh is uniform along the streamwise and spanwise directions, while a gradual mesh refining scheme was applied, in which the vertical mesh size becomes smaller in the approach to the air-water interface from either the water side or the air side (see figure 3). This meshing scheme was purposely used to resolve the air-water interface including centimeter-scale interfacial deformations as well as the molecular sub-layers in the air and water sides. The grid resolution for this mesh is between approximately 0.006 cm (near the air-water interface) and 0.05 cm (near the top and bottom of the domain).

Boundary conditions are summarized in figure 3b. The water side was started from rest with a flat air-water interface driven from above by the sudden imposition of a pre-computed, fully-developed boundary layer airflow driven by a pressure gradient (figure 4a). This boundary layer airflow was pre-computed (as a first step) in a separate simulation with a no-slip bottom and a zero-shear stress condition on top while maintaining periodic boundary conditions elsewhere (along streamwise and spanwise boundaries). The pre-computed airflow was characterized by far-field mean velocity $U_\infty = 5$ m/s, which is the same as the final far-field wind speed in the experiments of Veron and Melville [4]. For consistency, the computational mesh to run the preliminary airflow simulation was made identical to the air side portion of the air-water mesh in figure 3a.

The initial condition for scalar concentration was $C = 1$ mol/(m/s) in the air side and $C = 0$ in the water side. At the top of the domain (at the top of the air side) C was set to 1 mol/(m/s) and at the bottom of the domain (the bottom of the water side) C was set to 0, ensuring a flux of scalar from the air side to the water side (figure 3b).

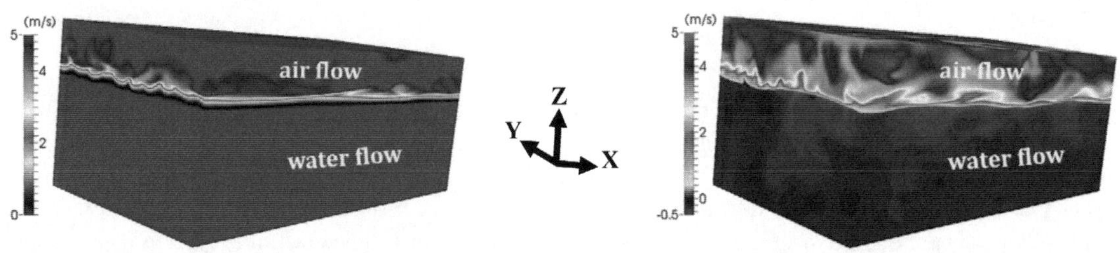

Figure 4. (a) Instantaneous snapshot of streamwise (x) velocity field distribution within the domain showing the turbulence in action within the air side and the water at rest as initial conditions. (b) Instantaneous snapshot of streamwise velocity field distribution within the domain showing turbulence in action within both the air and water sides after 5 s.

Figure 4b shows an instantaneous snapshot of streamwise (x) velocity field distribution within the domain showing the turbulence in action within both the air side and the water side at $t = 5$ s. Note the air flow velocities are much higher than the water flow velocities, as expected.

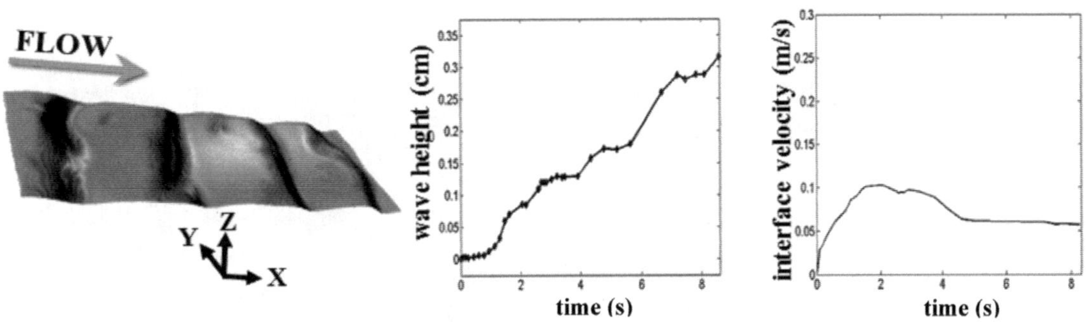

Figure 5. Left panel: the instantaneous air-water interface in DNS at $t = 2.5$ s. Middle panel: time series of the simulated maximum air-water interfacial wave heights. Right panel: time series of the simulated average streamwise velocity on the air-water interface.

Soon after the simulation started, at time $t = 1.5$ s, short capillary interfacial waves as small as 1 cm in wavelength and 0.25 mm in height were present accompanied by small capillary waves (ripples) with length scales of approximately 1–2 mm. The left panel in figure 5 shows a snapshot of the air-water interface deformation (corresponding to simulation time $t = 2.5$ s) of the DNS revealing gravity waves along with superimposed smaller capillary waves or ripples. The middle panel and right side panel of figure 5 show time series of maximum interfacial wave heights and average streamwise velocity on the interface, respectively. Trends observed in these records follow close to those observed in the DNS of Komori *et al.* [3].

As can be seen in figure 6a at $t = 0.5$ s, streamwise velocity fluctuations at the interface are characterized by smooth downwind elongated streaks. These narrow streaks are found to be parallel to each other, while alternating in sign in the spanwise direction. Starting at approximately $t = 1$ s (figure 6b), the streaky structure of the streamwise velocity fluctuation is disrupted by capillary wave deformations until about $t = 2.5$ s (figure 6c). This observed disruption is due to the fact that the downwind velocity fluctuation not only has a component due to the turbulence but it also has a significant capillary wave-induced component. As time progresses after $t = 2.5$ s, (e.g., at $t = 4$ s, figure 6d) the downwind elongated streaks re-emerge, due to the turbulent component of the velocity becoming more dominant than the wave-induced component. Re-emergence of downwind elongated streaks is indicative of the flow transitioning to Langmuir turbulence. Furthermore, the instantaneous scalar fluxes show a relative increase between $t = 2.5$ s (figure 6g) and $t = 4$ s (figure 6h). This increase is attributed to transition to Langmuir turbulence which contributes towards the enhancement of scalar uptake by the water. Note that the scalar flux is characterized by streaks similar to the velocity fluctuations.

Additionally, we compare results in terms of instantaneous streamwise vorticity in our simulation of the deforming air-water interface with instantaneous streamwise vorticity coming from a similar simulation but with the air-water interface held fixed (flat). The results from the flat interface case are shown on figures 7a and 7b and the results from the deforming interface case are shown on figures 7c and 7d. The deforming interface case is characterized by Langmuir cells growing in the cross-stream direction and in depth whereas the simulation with the flat interface shows smaller and less intense near-surface vortices.

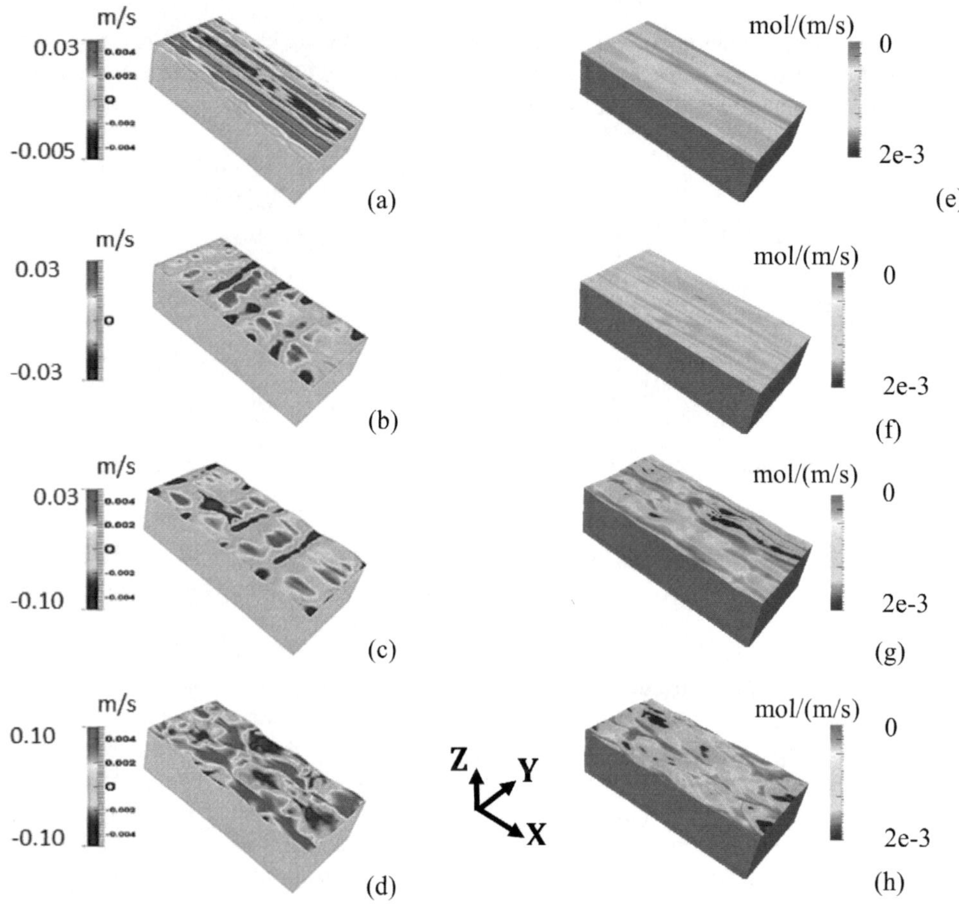

Figure 6. Instantaneous streamwise velocity fluctuation at (a) $t = 0.5$ s, (b) $t = 1$ s, (c) $t = 2.5$ s and (d) $t = 4$ s. Instantaneous scalar flux in mol/(m/s) at (e) $t = 0.5$ s and (f) $t = 1$ s, (g) $t = 2.5$ s and (h) $t = 4$ s.

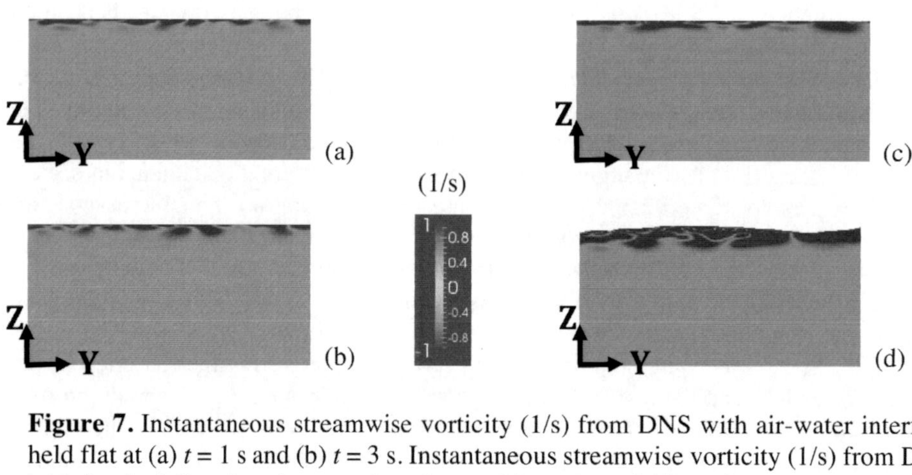

Figure 7. Instantaneous streamwise vorticity (1/s) from DNS with air-water interface held flat at (a) $t = 1$ s and (b) $t = 3$ s. Instantaneous streamwise vorticity (1/s) from DNS with freely deforming air-water interface at (c) $t = 1$s and (d) $t = 3$ s.

Moreover, prior to the transition to Langmuir turbulence, which was previously determined to be at approximately t = 2.5 s, both cases simulated (flat and deforming interface simulations) were compared in terms of depth profiles of mean scalar concentration in the water column. It was found that both cases possess similar concentration profiles through t = 2.5 s (not shown). However, for times greater than 2.5 s, the Langmuir turbulence and associated Langmuir cells in the deforming interface case generate greater vertical transport than the shear-generated turbulence in the flat interface case. From the depth profiles and instantaneous contours of scalar concentration shown in figure 8, we can conclude that the Langmuir turbulence penetrates deeper than the shear turbulence, thus the Langmuir turbulence is able to transport higher concentration fluid to greater depths. Note that as time progresses the difference between Langmuir turbulence penetration in the deforming interface case and the shear turbulence penetration in the flat interface case becomes more significant.

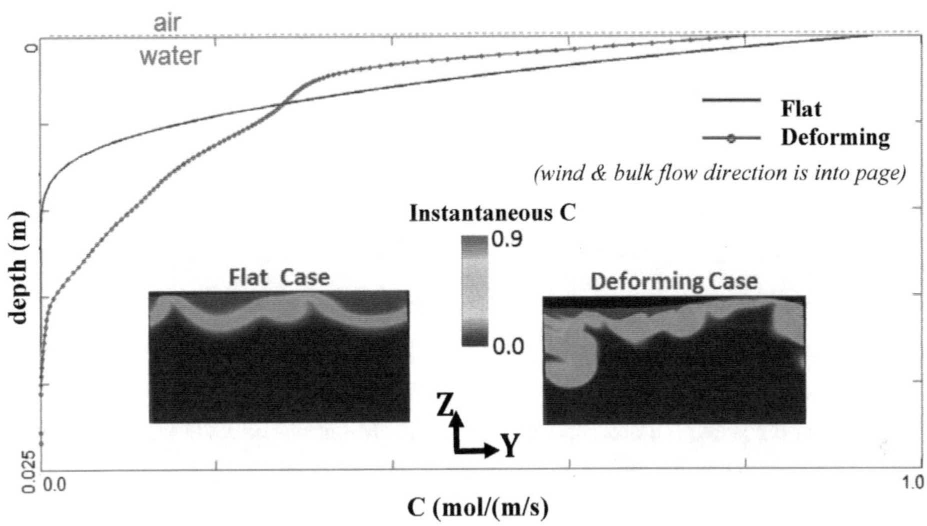

Figure 8. Depth profiles (averaged over streamwise and spanwise directions) and instantaneous snapshots of scalar concentration at time t = 4 s in the simulation with a deforming interface and the simulation with a flat interface.

The spatially averaged scalar flux across the wind-driven air-water interface for the flat and deforming interface cases are shown in figure 9. A dramatic explosion or spike of scalar flux is observed in the deforming case at approximately t = 2.5 s when the flow transitions to Langmuir turbulence. In contrast, a sudden increase in scalar flux is noticeably absent in the flat interface case. After this spike, the average scalar flux obtained in the deforming case decreases significantly but stabilizes at a mean value approximately two times greater than the scalar flux obtained in the flat case. In the field, it is likely that such gas flux spikes are correlated with wind transients or gusts and thus might be a dominant contributor to the long-term time-averaged gas flux.

Lastly, we calculate transfer velocity a measure of scalar transfer efficiency, following Kader *et al.* [1]:

$$k_L = \kappa \frac{\nabla C \cdot \boldsymbol{n}}{(C_i - C_{bulk})} \tag{4}$$

where C_i is streamwise and spanwise average concentration on the interface and C_{bulk} is bulk concentration calculated as

$$C_{bulk} = \int_0^{2\delta} \langle C \rangle_{x,y} \langle u \rangle_{x,y} dz \tag{5}$$

with $\langle \cdot \rangle_{x,y}$ denoting the average over streamwise and cross-stream directions. Due to the enhanced vertical transport induced by the Langmuir turbulence and associated cells, the bulk concentration is significantly greater in the interface deforming case compared to the flat interface case (figure 10). Ultimately this leads to a greater transfer velocity (k_L) in the deforming interface case (figure 11), as k_L is inversely proportional to the difference between concentration at the interface and bulk concentration (equation 4).

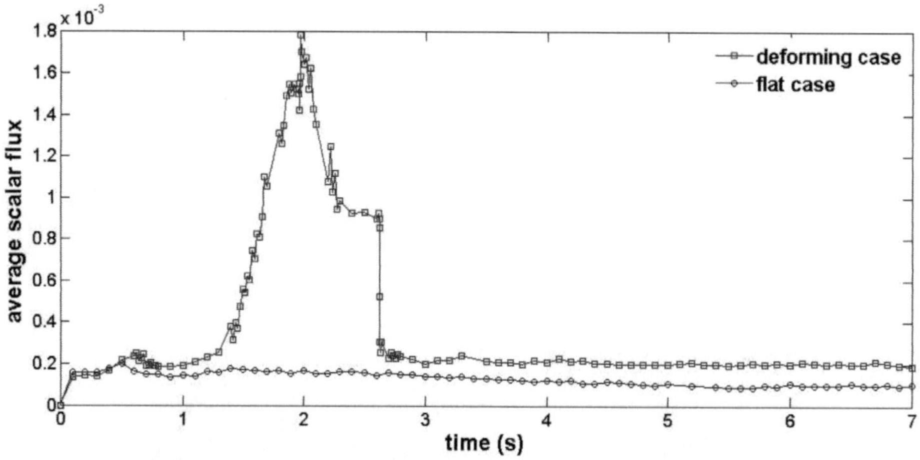

Figure 9. Average scalar flux through the air-water interface: $F = \kappa \nabla C \cdot \boldsymbol{n}$, where $\boldsymbol{n}$ is the unit normal to the interface.

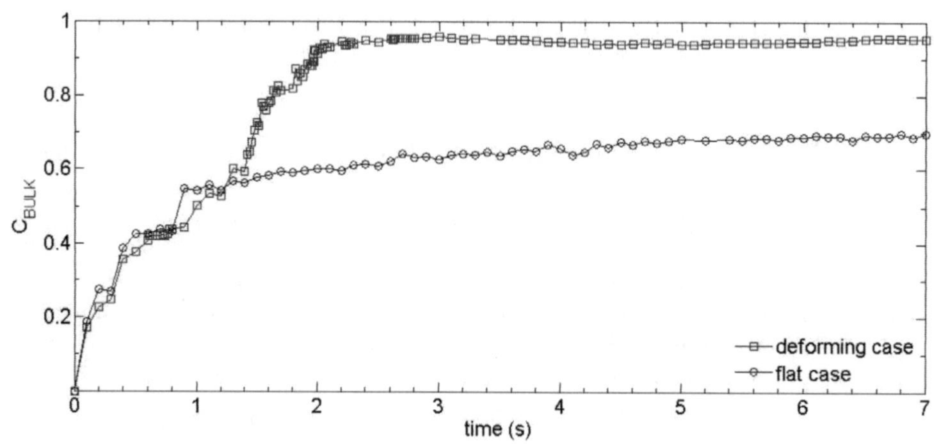

Figure 10. Bulk concentration of the air-water interface.

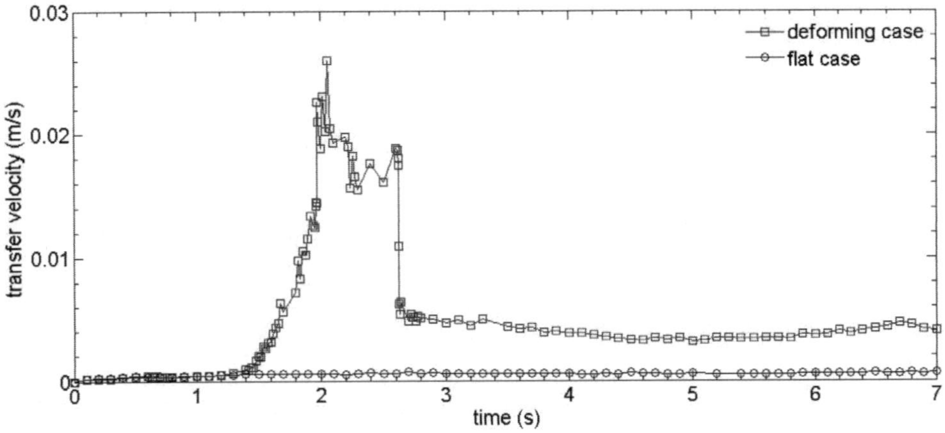

Figure 11. Transfer velocity.

4. Conclusion

The deforming interface and associated Langmuir turbulence play important roles in determining scalar flux across the interface and subsequent vertical transport of the scalar. Transition to Langmuir turbulence was observed to be accompanied by a spike in gas flux characterized by an order of magnitude increase. These episodic flux increases, if linked to gusts and unsteadiness in the wind field, are expected to be an important contributor in determining the long-term average of the air-sea fluxes. Finally, vertical transport induced by the Langmuir cells was seen to enhance bulk concentration throughout the water column which ultimately enhances transfer velocity. Future research will focus on understanding how the Langmuir turbulence affects scalar transfer at the interface via molecular diffusion.

References

[1] Kader B A and Yaglom A M 1972 *J. Heat Mass Transfer* **15** 2329
[2] Jasak H 1996 *Error Analysis and Estimation for the Finite Volume Method with Applications to Fluid Flows* (London: Imperial College)
[3] Komori S *et al* 2010 *J. Turbulence* **11** N32
[4] Veron F and Melville W K 2001 *J. Fluid. Mech.* **446** 25

Near-surface physics during convection affecting air-water gas transfer

S T Fredriksson[1], L Arneborg[1], H Nilsson[2] and R A Handler[3]
[1] Department of Marine Sciences, University of Gothenburg, P.O. Box 460, 405 30 Gothenburg, Sweden
[2] Department of Applied Mechanics, Fluid Dynamics, Chalmers University of Technology, 412 96 Gothenburg, Sweden
[3] Department of Mechanical Engineering, Texas A&M University, TX 77843-3123, USA

E-mail: sam.fredriksson@gu.se

Abstract. The gas flux at the water surface is affected by physical processes including turbulence from wind shear, microscale wave breaking, large-scale breaking, and convection due to heat loss at the surface. The main route in the parameterizations of the gas flux has been to use the wind speed as a proxy for the gas flux velocity, indirectly taking into account the dependency of the wind shear and the wave processes. The interest in the contributions from convection processes has increased as the gas flux from inland waters (with typically lower wind and sheltered conditions) now is believed to play a substantial role in the air-water gas flux budget. The gas flux is enhanced by convection through the mixing of the mixed layer as well as by decreasing the diffusive boundary layer thickness. The direct numerical simulations performed in this study are shown to be a valuable tool to enhance the understanding of this flow configuration often present in nature.

1. Introduction

The gas flux, F_g, across the air-water interface is often estimated as

$$F_g = k_g(C_w - \vartheta C_a), \quad (1)$$

where k_g is the gas transfer velocity, C_w and C_a are the gas concentrations in water and air, respectively (Figure 1a), and ϑ is the Ostwald solubility coefficient [1]. F_g is affected by several physical processes, e.g., interfacial shear due to wind forcing, microscale wave breaking at moderate wind speeds, bubbles, raindrops, breaking waves at high wind speeds, and convection due to surface heat loss [2]. k_g is frequently parameterized by empirical models based on the wind speed only. It is apparent that the wind speed, although contributing to many of these processes, is not enough to capture all the processes involved. This is especially true for low wind and zero-wind conditions where different wind speed parameterizations yield a large spectrum of gas transfer velocities ranging from zero upwards [1]. The gas flux during these conditions has received increasing interest since recent global model estimates suggest that inland waters, where low-wind conditions are more frequent than in the oceans, are considerable sources of greenhouse gases. The inland-water source, estimated to be 1 Pg C yr^{-1} [3-5], is of the same magnitude as the ocean net sink, which is estimated to be 1.6 Pg C yr^{-1} [4]. The present

Content from this work may be used under the terms of the Creative Commons Attribution 3.0 licence. Any further distribution of this work must maintain attribution to the author(s) and the title of the work, journal citation and DOI.
Published under licence by IOP Publishing Ltd

paper qualitatively discusses the gas flux dependency of natural convection during zero-wind conditions while Fredriksson et al. [6] present a more thorough method discussion and quantitative results.

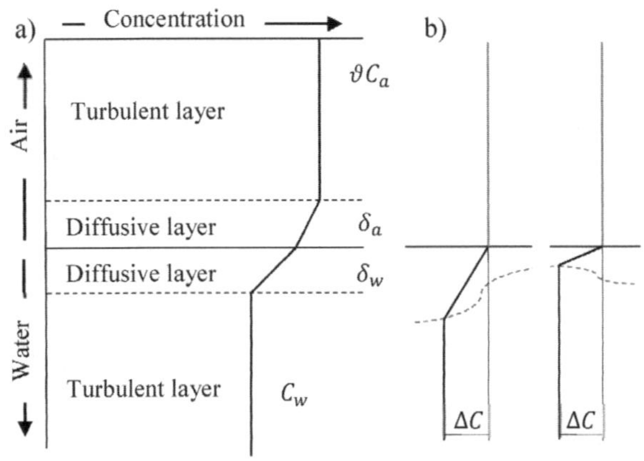

Figure 1. a) Conceptual diagram [1] of the mean gas concentration for a general gas without chemically enhanced diffusion in the water, the turbulent layers where advective transport dominates, and the boundary layers where diffusive transport dominates. b) Two snapshots of gas concentrations, assuming small gas concentration gradients in the air, and varying water diffusive boundary layer depths due to natural convection.

2. Modelling of air-water gas transfer

This study models the transport of a dissolved gas using a passive scalar in direct numerical simulations (DNS) of fully developed turbulent natural convective flow in the surface layer of oceans or lakes (Figure 2). The gas exchange across an air-water interface is affected by biochemical and physical factors. Biochemical factors, not considered in this study, are typically chemical or biological processes that produce or consume gas. Physical factors include advective and diffusive transport (Figure 1). For the water side, the advective motions dominate the transport of gas from the deeper water masses up to the diffusive sublayer, in which the vertical motions are attenuated and the molecular diffusion eventually takes over close to the surface. The actual gas exchange across the air-water interface is then, when bubbles or raindrops are not present, maintained by molecular diffusion driven by the air-water gas concentration gradient. The gas concentration gradients for CO_2 and CH_4 (typical greenhouse gases) are controlled by the water side [7], which allows studies of the exchange dynamics to be performed in the aqueous phase only. The horizontal gas concentration gradients for these gases are further assumed to be much smaller in the air than in water because the molecular diffusivities are several magnitudes higher in the air than in the water [8]. Hence a constant concentration boundary condition for the gas is imposed from the air side. The often used illustration of a constant gradient diffusive layer above a constant concentration layer (Figure 1a) is a rather coarse representation of the averaged quantities, not to speak about the momentary profiles. In reality the momentary concentration profiles show large variations of the diffusive layer thickness, and thereby also of the gradients and surface fluxes (Figure 1b). DNS is here used as a very valuable tool to study the momentary flow field sketched in Figure 1b in order to actually understand the processes that drive the flow and the gas transfer.

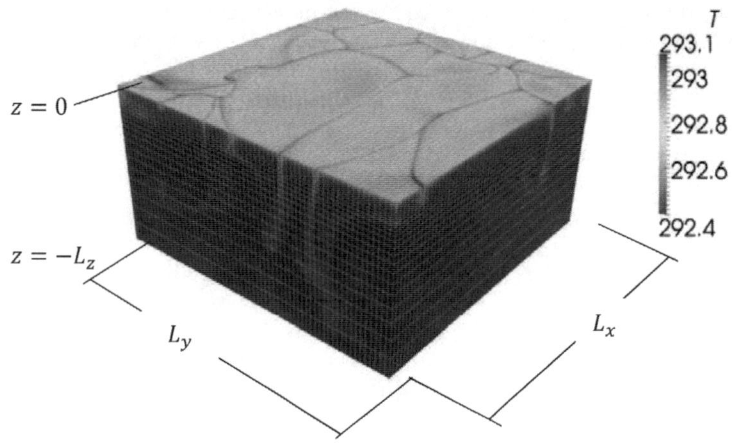

Figure 2. Computational domain, representing the surface layer of oceans or lakes. The mesh is equidistant in the x- and y-direction and graded in the z-direction. Every fourth mesh line in each direction is shown here. The constant surface heat flux cools the surface water and forms thin plumes of cold denser water moving downwards.

3. Governing equations, discretization, numerical cases, and statistical sampling

The flow is governed by the Navier-Stokes equations, which under the Boussinesq approximation [9] can be written as

$$\frac{\partial \boldsymbol{u}}{\partial t} + \boldsymbol{u} \cdot \nabla \boldsymbol{u} = -\frac{1}{\rho_0} \nabla P + \upsilon \nabla^2 \boldsymbol{u} + \beta (T - T_0) \boldsymbol{g}, \tag{2}$$

$$\nabla \cdot \boldsymbol{u} = 0, \tag{3}$$

and the thermal energy equation

$$\frac{\partial T}{\partial t} + \boldsymbol{u} \cdot \nabla T = \alpha \nabla^2 T + \phi_T. \tag{4}$$

Here $\boldsymbol{u} = (u, v, w)$ is the velocity vector where u, v are in the lateral directions and w is in the surface-normal direction. t is the time, υ is the kinematic viscosity of water, and P is a modified pressure with the background hydrostatic pressure for the constant density ρ_0 subtracted. β is the coefficient of thermal expansion, T is the temperature, T_0 is the reference temperature, and $\boldsymbol{g}$ is the gravitational acceleration. α is the thermal diffusivity and ϕ_T is a spatially and temporally constant source term added to maintain a constant mean temperature in the domain. The transport of the passive scalar s used to model the dissolved gas is governed by

$$\frac{\partial s}{\partial t} + \boldsymbol{u} \cdot \nabla s = D \nabla^2 s + \phi_s, \tag{5}$$

where $\phi_s = F_s / L_z$ is a spatially evenly distributed and temporally constant source term given by the scalar flux F_s through the top boundary and the domain depth L_z.

The domain is a non-deformable rectangular box, assuming that the deflection of the surface (top boundary) is negligible. The surface and bottom boundaries are assumed to be shear-free and the slip boundary condition $\partial u/\partial z = \partial v/\partial z = w = 0$ is applied on both. The surface heat flux is, during natural convection with no wind, dominated by long-wave radiation, which is assumed to be constant in the present simulations resulting in a constant gradient surface temperature boundary condition $\partial T/\partial z = -Q/\lambda$. Here $Q > 0$ is a constant heat flux directed out of the surface boundary, $\lambda = \rho c_p \alpha$ is the thermal conductivity, and c_p is the specific heat capacity. The scalar concentration at the surface is assumed to be constant $s = S_0$. The bottom boundary is assumed to be adiabatic and without scalar exchange with the lower water masses, i.e., $\partial T/\partial z = \partial s/\partial z = 0$. Finally, periodic boundary conditions are applied at the sides.

3.1. *Discretization*

The governing equations are solved using a collocated finite volume approach using the OpenFOAM open-source computational fluid dynamics code. The diffusion and advection terms are discretized using the second-order central differencing scheme and the time derivative is discretized using the Crank-Nicolson scheme [10]. The time step Δt is dynamically adjusted to keep the Courant-Friedrich-Lewy number (CFL) less than 0.5 in all cells. This typically gives a time step much smaller than the Kolmogorov time scale, and is therefore the limiting constraint for the time step.

The aspect ratios of the domain are $L_x = L_y = 2L_z$ and the mesh resolution is $256 \times 256 \times 96$ for all simulations. The mesh is graded in the vertical direction (Figure 2), where the distances between the center of the first cell to the bottom and surface boundary are 1.96 mm and 0.098 mm, respectively. The results presented in Fredriksson et al. [6] support the assumption that the current vertical and lateral dimensions of the domain are large enough to give representative results for the surface mixed layer although the dimensions are much smaller than in reality. The same study further concludes, via an extensive mesh resolution study, that the mesh resolution is appropriate to resolve the flow field at hand.

3.2. *Numerical simulations*

Three simulations are presented. The surface heat flux is varied as one *Base* case with a dimensional heat flux of 100 Wm^{-2} ($Ra = 5 \cdot 10^8$), as well as one *Low* case (50 Wm^{-2}, $Ra = 2.5 \cdot 10^8$), and one *High* case (200 Wm^{-2}, $Ra = 10 \cdot 10^8$). Here $Ra = \beta g Q L^4/\alpha v \lambda$ is the Rayleigh number using standard molecular properties for water at room temperature. The characteristic length, L, is here chosen as the domain depth $L_z = 0.1204\, m$. The Prandtl number, $Pr = v/\alpha$, and the Schmidt number, $Sc = v/D$, are 7 for all cases.

3.3. *Statistical sampling*

The sampling of the statistical properties of the turbulent fluctuations is done under fully developed conditions, defined by steady mean ($\bar{.}$) and root-mean-square (rms) for all presented variables. It is carried out during more than 40 (large eddy) outer time scales $t^* = L_z/W^*$, where $W^* = (BL_z)^{1/3}$ is the convective velocity and $B = \beta g Q/\rho c_p$ is the buoyancy flux. This corresponds to more than 500 inner time scales $t^+ = (v/B)^{1/2}$ [6]. The inner and outer scales refer to processes in the diffusive boundary layer and mixing layer, respectively. The flow is assumed to be periodic for all variables in the surface-parallel directions, resulting in statistical properties that only depend on the z direction. The horizontal time-space average during the sampling time t_A denoted $\langle . \rangle$ is defined by

$$\langle g \rangle(z) = \frac{1}{t_A}\int_t^{t+t_A}\left(\frac{1}{L_x L_y}\int_0^{L_x}\int_0^{L_y} g(x,y,z,t)\,dxdy\right)dt. \tag{6}$$

4. Results and discussion

4.1. *Enhanced scalar flux due to decreased diffusive boundary layer thickness*

Figure 3 shows a snapshot of the temperature field, the surface water velocities, the horizontal flow divergence, $\gamma = \partial u/\partial x + \partial v/\partial y$, and the surface-normal scalar transport, $F_s = -D\,\partial s/\partial z$, at the surface. It is shown in Figure 3a that the cooling of the surface creates cooler denser surface water that, due to instability, eventually forms thin descending plumes of cold denser water (sketched schematically with pale blue arrows). This cold water is replaced by warmer water ascending from the bulk in order to satisfy continuity. The descending cold water and the ascending warm water create the typical temperature field structures (Figures 2 and 3). It also causes straining (sketched schematically with yellow arrows) resulting in a thinner diffusive boundary layer and divergent and convergent regions at the surface. The thin diffusive boundary layer, δ_W, defined as where only 5% of the scalar transport is diffusive and the rest is advective, has been magnified 5 times in relation to the horizontal directions in

Figure 3b. The "islands" that can be seen to protrude the mean diffusive boundary layer, $\bar{\delta}_W \approx 5$ mm for the *Base* case, is the momentary diffusive boundary layer thickness. This momentary diffusive boundary layer thickness is also schematically drawn in red in Figure 1b. It is apparent in Figure 3b-c that there is a strong correlation between δ_W, the surface temperature, and the horizontal divergence. It is also shown in Figure 3b,d that wherever there is an "island" there is also an enhanced scalar flux through the surface, which is due to the increased scalar concentration gradient for thinner diffusive boundary layers.

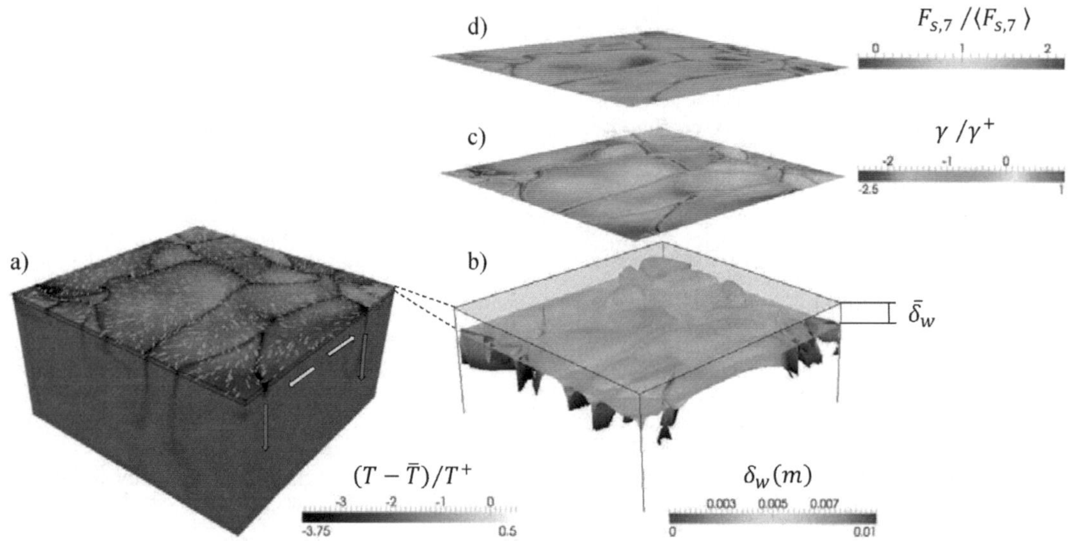

Figure 3. Snapshot, for the *Base* case, of a) the temperature field and the surface water velocities, b) the diffusive boundary layer depth, c) the horizontal flow divergence at the surface and d) the surface-normal scalar transport at the surface.

4.2. Different surface heat fluxes

Figure 4 shows the normalized temperature fields for different surface heat fluxes while keeping the domain size constant. It is interesting to find suitable normalization scales to describe the flow because these scales can be used to investigate, describe and predict other similar flow cases. Fredriksson et al. [6] discuss the use of inner and outer scales thoroughly, where the inner scales are used close to the surface, inside the diffusive boundary layer, and the outer scales are used in the mixed layer outside the diffusive boundary layer. Figure 4 shows that the scale of the thermal structures at the surface decreases with increasing heat flux as is expected according to the inner length scale $L^+ = Pr^{n-1}(\upsilon^3/B)^{1/4}$, where the buoyancy flux B is proportional to the heat flux. This length scale is further affected by Pr and the contamination of the surface, via the exponent n, typically between $1/2$ and $2/3$. The inner length scale is finally affected by the viscosity, but notably not the domain depth. The equivalent length scales for a scalar or a gas is $L^+ = Sc^{n-1}(\upsilon^3/B)^{1/4}$. Note, however, that the decrease in L^+ for gases with large Sc numbers is not reflected in the size of the temperature cells seen in Figure 4 but rather in the concentration gradients near the surface and the diffusive boundary layer thickness. The strong correlation between the inverse diffusive boundary layer thickness and the scalar flux seen in Figure 3 combined with the above expression for a typical vertical length scale leads to a scalar and gas transfer velocity that is proportional to $Q^{1/4}$, as confirmed by Fredriksson et al. [7].

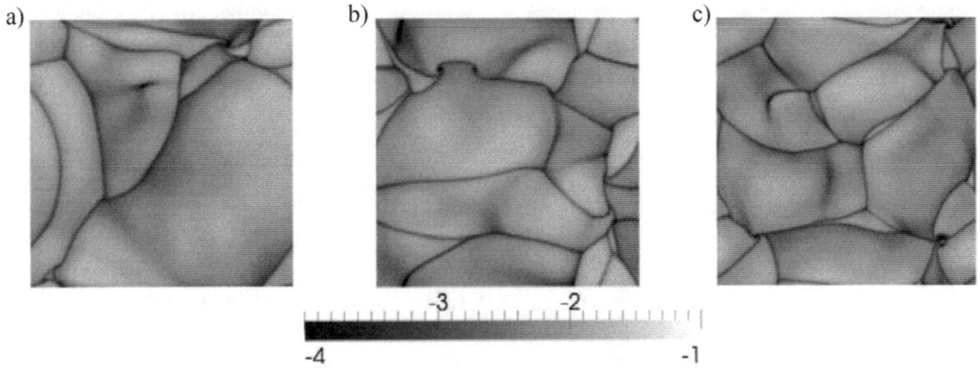

Figure 4. The normalised temperature field $(T - \bar{T})/T^+$ for *Low* a), *Base* b), and *High* c) surface heat flux, where the inner temperature scale $T^+ = QPr^n(Bv)^{-1/4}$.

4.3. Different temperature and scalar concentration profiles

The different boundary conditions for gas (modelled as a passive scalar) and temperature was presented in section 3. Figures 5 and 6 show the resulting concentration and temperature profiles in two different ways, where Figure 5 shows the minimum, mean, maximum, and momentary profiles along the y-z plane in the x-direction and Figure 6 shows the mean and the rms values for the complete sampling period normalised with the inner scaling T^+ and $s^+ = F_sSc^n(Bv)^{-1/4}$ [6]. It can be seen in Figure 5 that the scalar concentration is constant at the surface but the variation width increases with increasing depth while the variation width is fairly constant for the temperature all the way up to the surface. This is also shown in Figure 6 with s_{rms} and T_{rms} equal to zero and non-zero, respectively, at the surface. It can also be seen in Figure 6 that the scaled s_{rms} and T_{rms} become similar as the depth of the diffusive layer is approached. The gradient of the mean non-dimensional scalar concentration and temperature are similar, although the somewhat smaller gradient for temperature compared to the scalar concentration indicates that the transfer velocity is higher for the heat than the scalar.

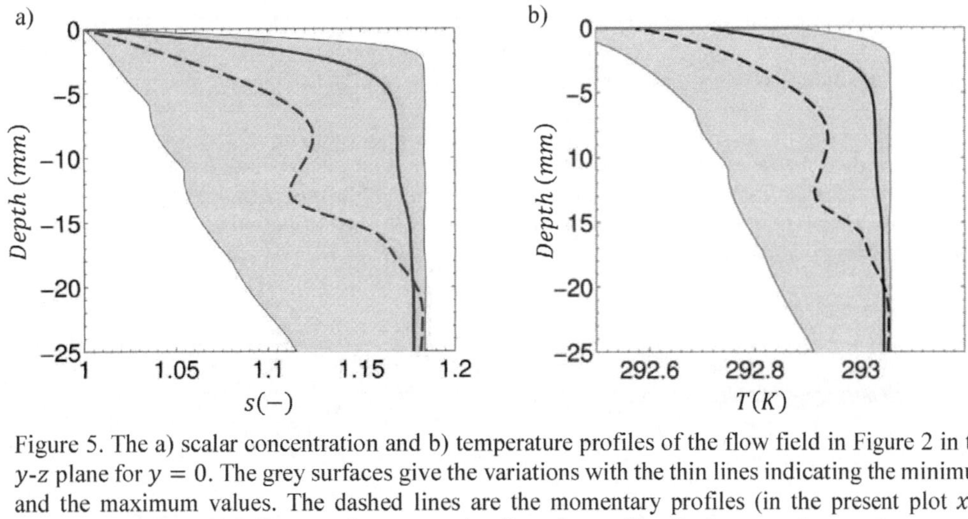

Figure 5. The a) scalar concentration and b) temperature profiles of the flow field in Figure 2 in the y-z plane for $y = 0$. The grey surfaces give the variations with the thin lines indicating the minimum and the maximum values. The dashed lines are the momentary profiles (in the present plot $x = L_x, y = 0$) and the thick lines are the mean value from the profiles for $0 < x < L_x, y = 0$.

The skewed distributions below the surface (Figure 5), with small minimum values compared to mean and max values, are caused by the descending plumes, and the increase in minimum scalar concentration and temperature with depth must therefore be caused by entrainment of ambient water into the plume.

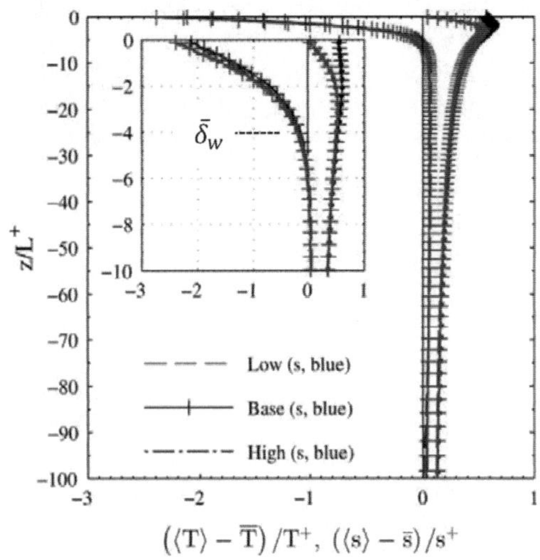

Figure 6. The mean and root-mean-square, normalized by inner scaling, of the fluctuating temperature, $(\langle T\rangle - \bar{T})$, and T_{rms} in black, and the fluctuating scalar concentration, $(\langle s\rangle - \bar{s})$, and s_{rms} in blue. The approximate mean diffusive boundary depth, $\bar{\delta}_w$, is indicated in the inset.

5. Summary

DNS is used to study gas flux across the air-water interface. The flow is driven by natural convection applied as a fixed surface heat flux and the gas transfer is modelled through a passive tracer (scalar concentration) transfer. The present paper discusses some of the results in the full DNS study presented in Fredriksson et al. [6], where the influence of a clean and a surfactant-laden surface as well as the Schmidt number dependence of the gas transfer velocity were studied. Fredriksson et al. [6] performed extensive mesh and domain size sensitivity analyses in order to verify that the results are robust and converged. In the case of a clean surface, modelled with a slip boundary condition, the buoyancy flux dependence was studied by altering the surface heat flux.

The DNS approach is shown to be powerful to enhance the understanding of the gas-transfer during natural convection. The use of inner scaling has been shown to synthesize the results in order to use the results for predicting similar flow cases. It is further illustrated that the momentary gas flux through the surface is closely related to the surface-normal diffusive boundary layer thickness, the temperature and the horizontal flow divergence field at the surface.

Acknowledgments

The computations were performed on resources provided by the Swedish National Infrastructure for Computing (SNIC) at C3SE (Chalmers Centre for Computational Science and Engineering) computing resources. We also acknowledge one anonymous reviewer.

IOP Conf. Series: Earth and Environmental Science **35** (2016) 012007 doi:10.1088/1755-1315/35/1/012007

References

[1] Bade D L 2009 Gas exchange at the air–water interface *Encyclopedia of Inland Waters* Likens G E, ed (Oxford: Academic Press) 70

[2] Macintyre S *et al* 2002 The critical importance of buoyancy flux for gas flux across the air-water interface *Gas Transfer at Water Surfaces* (American Geophysical Union) 135

[3] Bastviken D *et al* 2011 *Science* **331** 50

[4] Ciais P *et al* 2013 Carbon and other biogeochemical cycles *Climate Change 2013: The Physical Science Basis. Contribution of Working Group I to the Fifth Assessment Report of the Intergovernmental Panel on Climate Change* (Cambridge University Press)

[5] Tranvik L J *et al* 2009 *Limnol. Oceanogr.* **54** 2298

[6] Fredriksson S T *et al* 2015 An evaluation of gas transfer velocity parameterizations using DNS *J. Geophys. Res.* **Submitted**

[7] Jahne B and Haussecker H 1998 *Ann. Rev. Fluid Mech.* **30** 443

[8] Don W G and Robert H P 2008 Transport Properties *Perry's Chemical Engineers' Handbook, Eighth Edition* (McGraw Hill Profesional)

[9] Kundu P K *et al Fluid Mechanics, 5th ed* (Oxford, UK: Elsevier)

[10] Versteeg H K and Malalasekera W 2007 *An Introduction to Computational Fluid Dynamics: The Finite Volume Method, 2nd ed* (New York: Pearson Education Ltd)

A laboratory study of spray generation in high winds

D G Ortiz-Suslow, B K Haus, S Mehta and N J M Laxague
4600 Rickenbacker Causeway, Department of Ocean Sciences, Miami, FL 33149 USA
E-mail: dortiz-suslow@rsmas.miami.edu
E-mail: bhaus@rsmas.miami.edu

Abstract. Characterizing the vertical distribution of large spray particles (i.e., spume) in high wind conditions is necessary for better understanding of the development of the atmospheric boundary layer in extreme conditions. To this end a laboratory experiment was designed to observe the droplet concentration in the air above actively breaking waves. The experiments were carried out in hurricane force conditions (U_{10} equivalent wind speed of 36 to 54 m/s) and using both fresh water and salt water. While small differences between fresh and salt water were observed in profiles of radius-integrated spray volume fraction, the profiles tend to converge as the wind forcing increases. This supports the assumption that the physical mechanism for spume production is not sensitive to salinity and its corresponding link to the bubble size distribution.

1. Introduction

A respectable body of knowledge has been amassed regarding the open ocean exchange of momentum, energy, and gas across the ocean-atmosphere interface in light and moderate winds. However, information regarding these fluxes in high wind conditions remains limited. The atmospheric boundary layer becomes spray-ridden in wind speeds (referenced to a height of 10 m) above 30 m/s [1]. This has significant implications for modeling the balance between moist enthalpy input and wind energy dissipation at the air-sea interface and its impact on tropical storm development. It has been observed, in both laboratory [2] and field experiments [3, 4], that the creation of a defined spray layer coincides with the saturation of the atmospheric drag coefficient in the vicinity of 30 to 40 m/s. The fundamental mechanism that explains this remains elusive, but from a theoretical perspective it can be readily shown that the presence of spray in the bottom of the atmospheric boundary layer would alter the wind speed profile and subsequently the interfacial fluxes [5]. However, there is a dearth of direct knowledge regarding the effect that this increasing amount of entrained sea spray has on the air-sea flux of heat and momentum. Determining the relationship between sea spray and these transfer processes is critical for understanding the rate of intensification of tropical cyclones [6].

Particles with radii exceeding 25 µm are of particular interest as these have been shown to be the most significant to these processes [7, 8]. These spray droplets, also known as spume, are generated via wave breaking or wind shear tearing off the wave crests [9, 10]. Spume is generated in a range of sizes and it is expected that the presence of these large particles (50 µm–600 µm) will have an effect on the wind stress that is applied to the ocean surface. A number of theories exist that attempt to describe the role spray has on the transfer of momentum from the atmosphere to the ocean surface. Andreas [11] proposed that the falling droplets suppress the short wave development, which carry the majority of the wind stress. Also, while the wind forcing

Content from this work may be used under the terms of the Creative Commons Attribution 3.0 licence. Any further distribution of this work must maintain attribution to the author(s) and the title of the work, journal citation and DOI.
Published under licence by IOP Publishing Ltd

increases with wind speed, the amount of stress carried by the particles also increases and this effectively decreases the momentum transfer coefficient. An alternative theory is proposed by Barenblatt *et al.* and Bye and Jenkins [12, 13], who suggest that a spray-infested layer develops that essentially masks the ocean surface from direct physical interaction with the atmosphere. Soloviev and Lukas [9] suggest that due to the high wind shear the spray and spume production stabilizes the ocean surface and small waves become height-limited as their crests are blown off.

The presence of entrained spume particles can also act as an important thermodynamic layer between the atmosphere and ocean and is expected to affect the air-sea enthalpy flux. Taking into account the three-dimensional transport of a size-dependent distribution of aerodynamic particles that thermally interact with the ambient, turbulent flow is a challenge. Therefore, microphysical models proposed by Andreas [14], and expanded by Andreas and Emanuel [15], Andreas [16], and Andreas *et al.* [17], attempt to explain these dynamics in terms of the relevant time scales for droplet suspension, temperature evolution, and change in particle radius. A natural reference height for these models is the significant wave height, H_s, where the spray is expected to be generated at this level, travel through the atmosphere, and fall back to the surface where it can be considered coalesced with the ocean [17]. Jeong *et al.* [18] suggest that this physical description may change in strongly forced waves, where there is the potential for spray to impinge on the downwind wave slopes.

The role that spray and spume play in the fluxes across the air-sea interface all rely on some fundamental knowledge regarding the rate of particle production and the vertical distribution in the boundary layer. Accomplishing this in extreme conditions in the field is an arduous, if not impossible, task and so models and laboratories remain the primary sources of meaningful insight into these processes [19]. Currently, few laboratory experiments describe the vertical distribution of spume particles in either fresh or salt water in the strong wind forcings relevant to extreme oceanic conditions.

The results of a laboratory experiment focused on obtaining direct observations of the vertical distribution of large spume particles in very high winds are presented here. A nonintrusive, optical method is used to image spray droplets at a range of heights above the mean water level (MWL) in U_{10} equivalent winds ranging from 36 to 54 m/s. The particles observed range in radii from 86 µm to 1386 µm and experimental trials were conducted in both fresh and salt water. It is well known that bubble production in salt and fresh water exhibits distinct differences in the size and number of bubbles produced. While this is likely to have a large impact on spray aerosol production through the bubble bursting mechanism, it is not known whether this will significantly impact spume production. Here we seek to determine if there are significant differences in spume production for salt and fresh water. This will be important for comparing datasets collected over water bodies with different salinities.

2. Data collection

The observations were made in the University of Miami Air-Sea Interaction Saltwater Tank (ASIST), which is a wind-wave-current flume with a 15 x 1 x 1 m acrylic test section (see figure 1). Spray observations were made 11.05 m downwind of the air inlet. Maximum sustained winds, as measured by a sonic anemometer 6.6 m downwind of the inlet, reached ~54 m/s when referenced to 10 m. This scaling follows similar work in ASIST by Donelan *et al.* [2] and Haus *et al.* [20]. A reference water depth of 0.42 m was used for all of the experiments.

Spray droplets were imaged using a modified Dantec Dynamics PIV (particle image velocimetry) acquisition system, where the laser sheet was re-routed through a liquid light guide to a strobe equipped with a telecentric lens, which produces a columnar beam of light (i.e., essentially non-convergent) that enhances the image contrast as opposed to a standard diffuse light source. This technique is similar to the direct imaging technique described in Christensen and Thomasson [21] and has been shown to be a robust and relatively straightforward method for

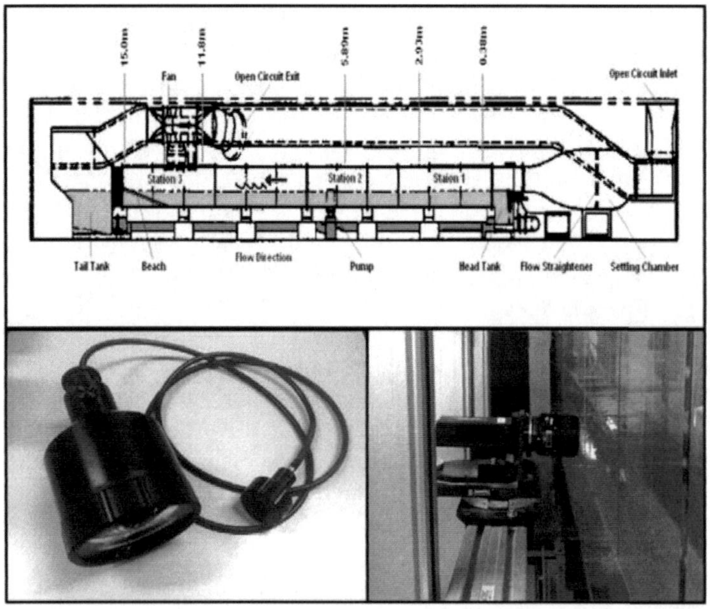

Figure 1. Clockwise from top: a schematic of the ASIST; the camera used for the data collection shown here on the vertical traverse that enabled multi-level imaging; the strobe and liquid light guide, which was mounted on another traverse (not shown) opposite the camera. The green tinge in the water is from a fluorescent dye, the observations presented here were done with clean, non-dyed water for both fresh and saline conditions.

quantifying particle statistics. This method differs from other optical techniques used in similar-style laboratory experiments [22], such as phase Doppler anemometry, which use convergent light sources to back-light particles and require more extensive cross-sectional area calibrations [23].

The camera used here (JAI CV-MSCL, 1.9 MP, 30 fps) was equipped with a medium telephoto lens (field of view 23.3°). This lens exhibits a magnification (pixels per meter) dependence on the depth of field; however, in practice this potential source of error is relatively small. Given that the focal plane was calibrated at 0.59 m from the lens (i.e., this is the focusing distance approximately at the center line of ASIST) and the depth of field of this lens is of order 3 mm, the potential magnification error in the particle sizing is around 1%. This source of error is much less than the uncertainties in the particle sizing algorithm used for this study, discussed further in section 3.

The collimated light beam was directed onto a diffusing screen (to reduce intensity) and the camera was oriented opposite the light source such that the optical axis of both strobe and camera were colinear. This camera-strobe system (including screen) was oriented perpendicular to the test region and mounted outside of the acrylic walls of ASIST. This meant that any spray droplets passing between light source and camera appeared as two-dimensional shadow projections (or silhouettes). The advantage of using this imaging method was that the images could be calibrated at a single focal plane [21], which was located in the middle of the ASIST control volume and oriented in a plane perpendicular to the along-tank direction. This experimental set-up enabled 42 µm per pixel resolution.

Five trials were conducted with wind speeds ranging in U_{10} equivalent magnitude from 36 to 54 m/s. For each trial, the tank was allowed 120 s to achieve steady state and then several image collections were acquired continuously, which generally took about 175 s to complete. For

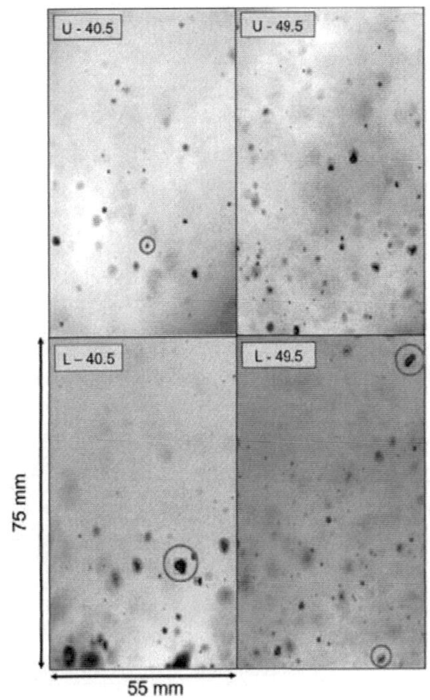

Figure 2. Sample images from the spray imaging data set. The U (L) refers to the upper (lower) level images and the number refers to the U_{10} equivalent wind speed. The red boxes and yellow circles represent particles identified and contoured by the automatic processing algorithm. Clockwise from upper left, the red-circled droplets have area equivalent radii of 92.5, 303.5, 181.5, and 396 µm.

each collection a total of 250 image pairs (i.e., 500 total frames) were acquired. The timing is precisely controlled by the Dantec software, and the frames within a pair were separated by 500 µm, with respective pairs sampled at 66.667 milliseconds (or 15 Hz). These five wind trials were repeated for both fresh and salt water. The salt water was 2 µM filtered sea water pumped in from a nearby tidal inlet, which can be considered a fully marine environment with no nearby fresh water sources.

In order to capture the vertical variability, each wind trial was sampled at different heights above the MWL. The salt water data were sampled at two reference heights, with the image frames centered at 95 mm (lower level) and 145 mm (upper level) above the mean surface. The fresh water data were sampled at three vertical levels, with frames centered at 95 mm, 145 mm, and 195 mm, respectively. Each individual experimental trial (i.e., water condition, a given wind speed, and reference level) was conducted independently with initial conditions in the tank being restored before another data set was collected. In total, this spray data set represents over 70,000 individual frames.

3. Data processing

The images were processed with a Dantec Dynamics shadow imaging software package using a two-step procedure. While image pairs were acquired, only one image per pair was used to compute the particle statistics. In the early stages of the image processing it was discovered that the sampling interval between pairs was too large to properly resolve the particle advection reliably across the entire data set, but single images were sufficient to determine droplet statistics. In the first step, each raw frame was balanced in order to correct for irregularities in the background light intensity of the imaged area. A collection of 250 frames was analyzed to determine the mean light sheet, then individual frames from this set of 250 were balanced according to this mean. Because this analysis was applied adaptively to each set independently, the overall light intensity reduction as the images became filled with droplets at the highest wind speeds was automatically taken into account. This helped enhance the image contrast across the entire data set and normalize the images towards a standard gray level contrast between the background and the particles. A droplet identification algorithm was then applied to these now-balanced frames. The algorithm was trained on a single frame in order to determine baseline gray level contrast and particle edge gradients. The results of this training were then applied to the entire set of 250 images. This procedure was repeated independently for each respective collection of 250 frames (i.e, given a water state, wind speed, etc.).

This semi-automatic algorithm allowed this extensive data set to be processed far more expeditiously and consistently than if it had been manually analyzed. An attempt was made to quantify the success rate of this method and provide some confidence in its skill as a first step in post-processing quality control. For each 250-image collection, every 25th frame was visually inspected in order to ground truth the automatic detection algorithm. The automatic processing successfully identified (and sized) 75–90% of the droplets for all the data except those in lowest wind speed trial, which was only able to identify just above 60%. This discrepancy can be reasonably explained by low image contrast and variability in the light sheet that could not be adequately balanced in the processing. Also, a radius dependence was observed in the success rate, with the algorithm far more likely to miss the smaller particles (a few pixels across in diameter) as opposed to the larger drops (these were nearly always identified correctly). This was done for both experiments. The algorithm was observed to work significantly better with the later data set. This discrepancy in data quality could exist for various reasons due to variable image collection conditions that could not be rectified via the processing algorithm. The results presented in section 4 only include fresh water trials that were deemed acceptable given the results of the user-verifications.

4. Results

Spray concentration was observed to increase with wind speed and decrease with height above the surface. This was quantified following [24] as the number of particles per unit air volume per discrete radius interval,

$$n(r_i, z_j) = \frac{Count(r_i, z_j)}{a(z_j)U dt dr}, \tag{1}$$

where i and j are simply indices. The $Count(r_i, z_j)$ is the total number of particles for a particular vertical bin and radius class, $a(z_j)$ is the cross-sectional area of the vertical bin, U is the wind speed, dt and dr are the time interval and radius increment (here 50 µm), respectively. The along-tank and vertical extent of the total sampling volume was 55 mm and 75 mm, respectively, with a corresponding across-tank extent of 70 mm. This across-tank extent was determined in post-calibration as the operational depth of field. This was quantified using a standardized target with circles of known diameters between 1 and 2.5 mm. Investigations into the particle detection algorithm results revealed that the edge-detection method could not discriminate out-of-focus particles to the precision of the depth of field of the equipped lens (i.e., ± 1.5 mm). Thus the across-tank dimension used here can be considered the effective region sampled within each image by the particle detection method. By examining the grayscale gradient on the edge of out-of-focus standardized circles, this across-tank dimension (i.e., operational depth of field) was found to be largely size-independent. The vertical dimension used to estimate the air volume was 3 mm or the resolution of the profile.

The $n(r, z)$ for the highest (54 m/s) and lowest (36 m/s) wind conditions in both fresh and salt water (figure 3) provide a two-dimensional view of the droplet concentrations and show the filling of the laboratory air boundary layer with spume particles. For all particle sizes, the number concentration tends to decrease with height above the surface and increase with wind speed. The vertical gradients in number concentration for the smallest observed particles <200 µm tend to reduce with increasing wind speed, as in the z dependence becomes negligible.

Table 1. Percent of the $n(r, z)$ that was observed to be "empty" for fresh and salt water in all wind speed conditions considered in this study. Examples of the distributions are given in figure 3.

	Wind Speed			
Water Type	36 m/s	40.5 m/s	49.5 m/s	54 m/s
fresh	85.3	70.6	53.3	44.9
salt	84.5	65.5	37.0	32.4

However, this never occurs for the particle radii approaching 1 mm as their strong vertical gradients persist even during the strongest wind forcing. The height above the MWL here is scaled by the respective H_s,

$$H_s = 4\sqrt{m_0}, \tag{2}$$

where m_0 is the 0th moment of the surface elevation variance spectrum derived from time series collected via an Ultrasonic Distance Meter (UDM), which was sampled at 10 Hz and located 5 m upwind of the imaging system used here.

In comparing fresh water to salt water, the distributions in figure 3 appear qualitatively similar, but some subtle differences exist, which warrant highlighting. The number concentrations are presented here as a two-dimensional particle density distribution that is a

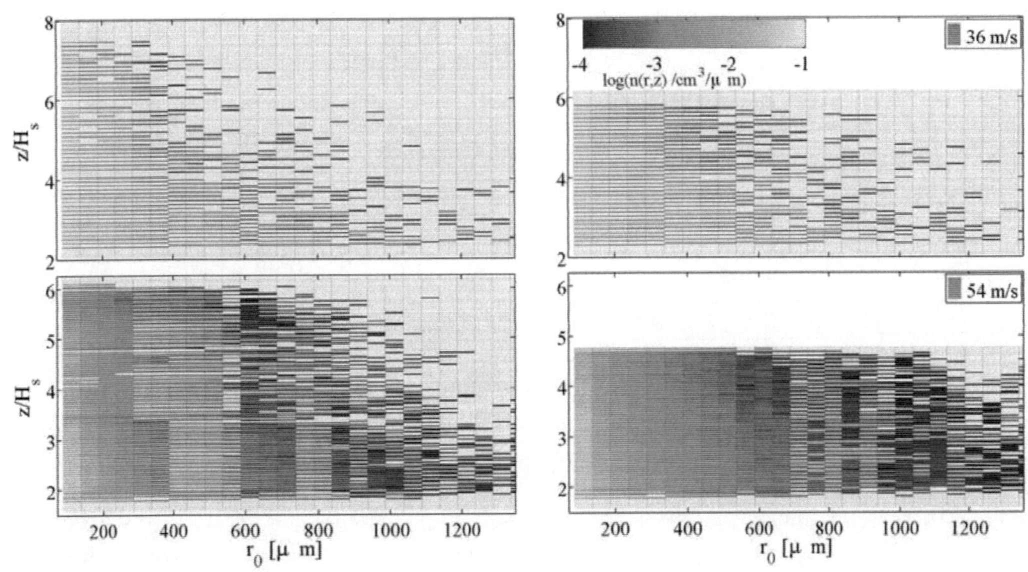

Figure 3. Two-dimensional distributions of number concentration as a function of height and observed particle radius for fresh (left column) and salt (right column) water. The lowest wind speed (upper row) and highest wind speed (lower row) trials are provided for comparison. Color refers to the log-scaled number of particles per unit air volume per radius class, the color scales are equivalent across the four panels. Grayed cells represent no particles counted, white regions represent unsampled physical space.

function of z/H_s and r (the latter is the observed radius). For fresh water, this distribution is concentrated in the small particle regime of the observed spectrum, while for saline conditions this distribution is far broader. This is evidenced by the relatively higher number concentrations observed in salt water opposed to fresh water out to an observed radius near 600 µm. This is seen at 36 m/s, but becomes more obvious in the 54 m/s wind trials (figure 3).

This may be quantified by the "emptiness" of the distribution. Given the discretization done in the averaging, there are portions of these arrays where no particles are actually counted for a given (z/H_s,r) coordinate (this information in terms of percentage for both water states is given in table 1). These percentages were only calculated over the shared vertical extent between the two data sets. There is a clear trend of decreasing emptiness with increasing wind speed, but this transition occurs faster in the saline water. From the distributions it is evident that this filling occurs most significantly in the larger particle regime. This may provide some evidence of the limiting effect of evaporation in the salt water, where particles can only lose a certain amount of volume during their aerial transport [24].

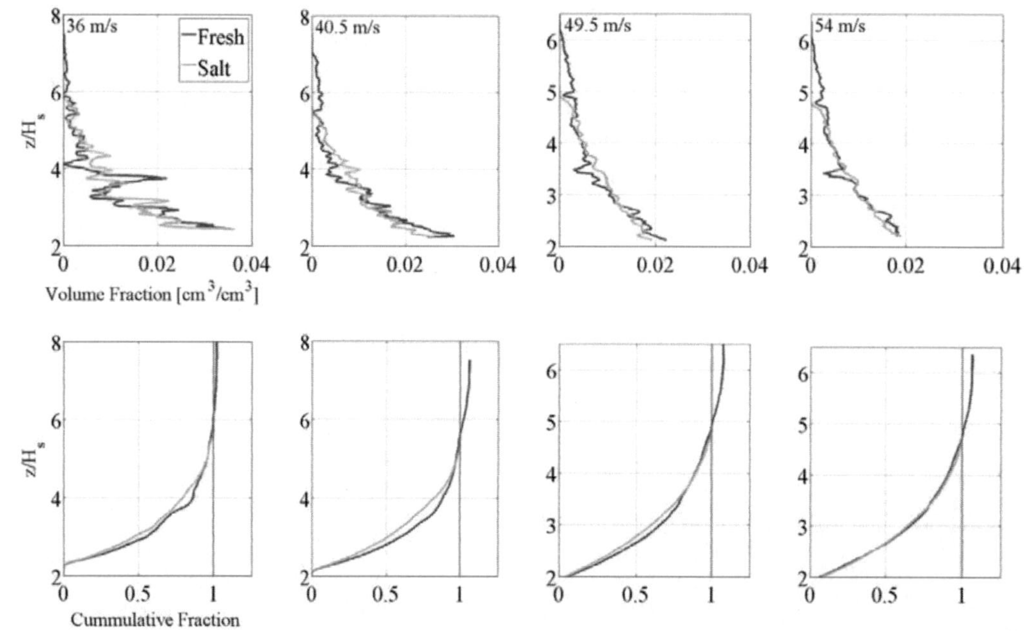

Figure 4. (Upper row) The total spray volume fraction profile for fresh and salt water across all wind speed conditions. (Lower row) Corresponding cumulative summation. Note, the fresh water was normalized up to the top of the salt water profile in order to make a fair comparison across data sets—this explains the >1 cumulation.

To highlight the vertical dependence, the radius-integrated $n(r, z)$ can be scaled into spray volume fraction profiles, as in the volume of spray observed in a vertical bin per total spray volume observed across the entire profile (figure 4). In order to reduce noise due to the discretization, these profiles have been smoothed with a running median filter two vertical bins wide. The profiles for fresh and salt water are generally similar in shape, but consistently in fresh water more spray volume is lost vertically than is observed to occur in salt water. This can also be seen in cumulative summations where the salt water profiles take longer than the fresh water to reach an equivalent fraction. These differences are fairly small and the profiles converge with increased forcing.

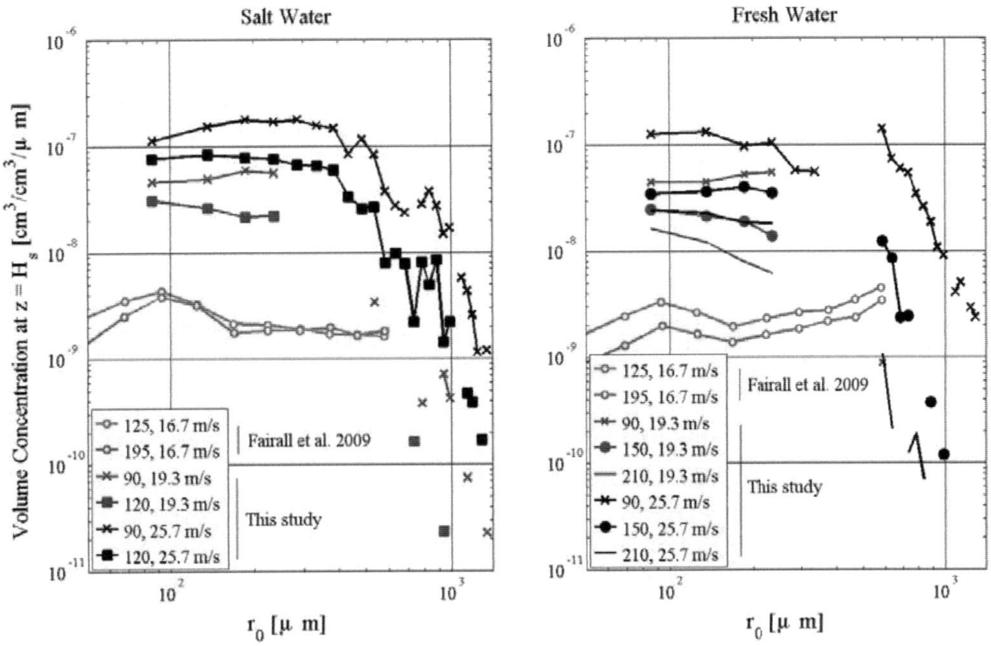

Figure 5. Size-dependent spray volume concentrations transformed to the theoretical source level ($z = H_s$) for salt (left) and fresh (right) water, as in the volume of spray per unit volume of air per radius increment. The legend entries denote the height above the MWL of the original observation point in mm and the wind forcing in m/s. Note that these wind speeds are in a lab reference frame, as opposed to the 10-m winds referenced throughout this work. This is done to make fair comparisons to Fairall *et al.* [24] (magenta and red curves in both panels), who did not report scaled wind speeds. 25.7 m/s and 19.3 m/s correspond to the 54 m/s and 40.5 m/s wind speed trials reported within this study, respectively.

The number concentrations from equation (1) can be converted to volume concentrations by multiplying by $4/3\pi r_i^3$ and transforming down to the theoretical source height, H_s, using methods described in Fairall *et al.* [24] (figure 5). It is expected that for a given wind speed and radius class the volume concentrations observed at various heights above this level should all collapse to the "source" value. A convergence within 0.5 orders of magnitude is observed, which is slightly more than reported by Fairall *et al.* [24], but is nonetheless fairly consistent with accepted theories. It should be noted that the Fairall *et al.* [24] study used 24 ppt water to simulate oceanic salinities, which differs from the present work that utilized filtered sea water (32–34 ppt) for the saline trials. Also, the spectra from that study were smoothed to fill in gaps while the curves from the observations presented here were not.

With regards to the ASIST data, fresh and salt water spectra are remarkably similar. This does not agree with observations reported by Fairall *et al.* [24], where fresh and salt water show markedly different behavior for particles >200 μm (see figure 5). In that work these differences were attributed largely to evaporation as well as a potential sampling bias. The wind forcing dependence across water salinities is markedly similar, which is in agreement with previous observations [24]. These findings provide indirect confirmation that bubble production is not an important spume generation mechanism, because bubbles are known to be larger in fresh water versus salt water.

5. Future work

This study has attempted to characterize the vertical distribution of spume particles observed above breaking waves in both fresh and salt water states in a series of laboratory experiments. To first order salinity seems to play little role in the production of spume particles in high winds; however, differences in both the vertical profiles and particle radius spectrum were noted. Further investigation into the affect that this has on the vertical spray flux will be the focus of future efforts.

Acknowledgments

This work is supported by the NSF through grant 0933943. The authors appreciate the efforts of Mike Rebozo and Neil Williams for their help with the data collection. Professor M. Donelan is thanked for the acquisition of and permission to use previously collected data in ASIST.

References

[1] Fairall C, Kepert J and Holland G 1994 *Glob. Atmos. Ocean Syst.* **2** 121–142
[2] Donelan M A, Haus B K, Reul N, Plant W J, Stiassnie M, Graber H C, Brown O B and Saltzman E S 2004 *Geophys. Res. Lett.* **31** 5
[3] Powell M D, Vickery P J and Reinhold T A 2003 *Nature* **422** 279–83 ISSN 0028-0836
[4] Potter H, Graber H C, Williams N J, Collins C O, Ramos R J and Drennan W M 2015 *J. Atmos. Sci.* **72** 104–118 ISSN 0022-4928
[5] Pielke R A and Lee T J 1991 *Boundary-Layer Meteorol.* **55** 305–308 ISSN 0006-8314 URL `http://link.springer.com/10.1007/BF00122582`
[6] Soloviev A V, Lukas R, Donelan M A, Haus B K and Ginis I 2014 *Sci Rep* **4** 5306 URL `http://www.ncbi.nlm.nih.gov/pubmed/24930493`
[7] Lewis E R and Schwartz S E 2004 *Sea Salt Aerosol Production: Mechanisms, Methods, Measurements, and Models - A Critical Review* (American Geophysical Union) ISBN 0875904173
[8] Jones K F and Andreas E L 2012 *Q. J. R. Meteorol. Soc.* **138** 131–144 ISSN 00359009
[9] Soloviev A and Lukas R 2010 *Boundary-Layer Meteorol.* **136** 365–376 ISSN 00068314
[10] Anguelova M, Barber R P and Wu J 1999 *J. Phys. Oceanogr.* **29** 1156–1165 ISSN 0022-3670
[11] Andreas E L 2004 *J. Phys. Oceanogr.* **34** 1429–1440 ISSN 0022-3670
[12] Barenblatt G I, Chorin A J and Prostokishin V M 2005 *Proc. Natl. Acad. Sci. U. S. A.* **102** 11148–11150 ISSN 0027-8424
[13] Bye J a T and Jenkins A D 2006 *J. Geophys. Res.* **111** 1–9 ISSN 0148-0227
[14] Andreas E L 1992 *J. Geophys. Res.* **99** 14345 ISSN 0148-0227
[15] Andreas E L and Emanuel K a 2001 *J. Atmos. Sci.* **58** 3741–3751 ISSN 0022-4928
[16] Andreas E L 2005 *Atmos. Res.* **75** 323–345 ISSN 01698095
[17] Andreas E L, Jones K F and Fairall C W 2010 *J. Geophys. Res. Ocean.* **115** 1–16 ISSN 21699291
[18] Jeong D, Haus B K and Donelan M A 2012 *J. Atmos. Sci.* **69** 2733–2748 ISSN 0022-4928
[19] Melville W K 1996 *Annu. Rev. Fluid Mech.* **28** 279–321
[20] Haus B K, Jeong D, Donelan M A, Zhang J A and Savelyev I 2010 *Geophys. Res. Lett.* **37** n/a–n/a ISSN 00948276
[21] Christensen M and Thomassen P 2014 *Experimental Determination of Bubble Size Distribution in a Water Column by Interferometric Particle Imaging and Telecentric Direct Image Method* Masters thesis Aalborg University
[22] Asher W and Farley P 1995 *J. Geophys. Res.* **100** 7045–7056
[23] Saffman M 1987 *Appl. Opt.* **26**
[24] Fairall C W, Banner M L, Peirson W L, Asher W and Morison R P 2009 *J. Geophys. Res.* **114** C10001 ISSN 0148-0227 URL `http://doi.wiley.com/10.1029/2008JC004918`

Estimation of friction velocity from the wind-wave spectrum at extremely high wind speeds

N Takagaki[1], S Komori[1] and N Suzuki[2]

[1]Department of Mechanical Engineering and Science and Advanced Research Institute of Fluid Science and Engineering, Kyoto University, Kyoto 615-8540, Japan
[2]Department of Mechanical Engineering, Kinki University, Osaka 577-8502, Japan

komori@mech.kyoto-u.ac.jp

Abstract. The equilibrium range of wind-waves at normal and extremely high wind speeds was investigated experimentally using a high-speed wind-wave tank together with field measurements at normal wind speeds. Water level fluctuations at normal and extremely high wind speeds were measured with resistance-type wave gauges, and the wind-wave spectrum and significant phase velocity were calculated. The equilibrium range constant was estimated from the wind-wave spectrum and showed the strong relationship with inverse wave age at normal and extremely high wind speeds. Using the strong relation between the equilibrium range constant and inverse wave age, a new method for estimating the wind speed at 10-m height (U_{10}) and friction velocity (u*) was proposed. The results suggest that U_{10} and u* can be estimated from wave measurements alone at extremely high wind speeds in oceans under tropical cyclones.

1. Introduction

It is of great importance to predict the development and decay of tropical cyclones accurately in order to minimize loss of life and property, since tropical cyclones can wreak catastrophic damage to a local society by generating gales, storm surges, and heavy rain. Reliable predictions of tropical cyclone intensity require accurate estimates of momentum and heat transfer across the sea surface. An important goal in constructing a reliable model to forecast the decay or development of a tropical cyclone is how to accurately estimate momentum transfer rate across the air-sea interface in stormy conditions. Such an estimate requires an accurate drag coefficient (C_D) model against wind speed at z = 10 m above the sea surface (U_{10}). The sea surface wind shear stress (τ) is expressed as:

$$\tau = \rho u^{*2} = \rho C_D U_{10}^2, \tag{1}$$

where ρ is the density of air and u* is the air friction velocity.

Recent field measurements [1] made by a Global Positioning System dropsonde in a tropical cyclone showed that C_D decreases with increasing U_{10} at extremely high wind speeds of $U_{10} \gtrsim 35$ m/s, whereas previous field [2] and laboratory measurements [3] suggested that C_D increases with increasing U_{10} also at extremely high wind speeds. Recent laboratory data [4, 5] show almost constant values of C_D at extremely high wind speeds of U_{10} = 35 ~ 68 m/s; data reported by Takagaki *et al.* [5] were measured directly by an eddy correlation method [see 6, 7] using a phase Doppler anemometer (PDA). Thus, the trend of C_D at extremely high wind speeds remains controversial: increasing, decreasing, or constant

Content from this work may be used under the terms of the Creative Commons Attribution 3.0 licence. Any further distribution of this work must maintain attribution to the author(s) and the title of the work, journal citation and DOI.
Published under licence by IOP Publishing Ltd

with U_{10}. For field measurements, very expensive dropsondes and aircraft are used to measure wind speed in and under tropical cyclones. These dropsondes have revealed significant new evidence for momentum transfer across the sea surface under tropical cyclones [e.g., 1, 8, and 9]. However, it is prohibitively expensive to use dropsondes and aircraft to observe entire tropical cyclones and forecast weather. Therefore, alternative simple instruments and techniques for measuring U_{10} and C_D under tropical cyclones are needed.

At normal wind speeds of U_{10} < 35 m/s in the absence of tropical cyclones, spaceborne radar observations are generally used to estimate wind speeds near the sea surface by exploiting the strong relationship between the small-scale surface structure on wind-waves and the relative backscattering intensity of radar. For example, previous studies [10–13] developed several methods to estimate U_{10} or $u*$ from wind-wave data. The mean square slope of the sea surface [10, 11] or the equilibrium range of the wind-wave spectrum [12, 13], which gives wind-wave energy level, are used to estimate wind speeds at U_{10} < 20 m/s. And Donelan *et al.* [4] demonstrated that general 5.3 GHz (C band) microwave radar observations are not applicable to observations under tropical cyclones. However, if an equilibrium range exists at extremely high wind speeds, simple physical-based instruments and techniques for measuring U_{10} and C_D under tropical cyclones might be feasible.

The purpose of this study is, therefore, to clarify the wind-wave spectral shape, then to clarify the existence of an equilibrium range of wind-waves at extremely high wind speeds, and to develop a model for calculating the dimensionless equilibrium range constant. By using the model for the dimensionless equilibrium range constant, we propose a method to estimate U_{10} (or $u*$) and quantify the error of this method.

2. Experiments

2.1. Laboratory experiment

The high-speed wind-wave tank [5, 14, and 15] with a glass test section 15.0 m long, 0.8 m wide, and 1.6 m high was used. Wind-waves were driven in the water tank by wind at U_{10} = 7 – 68 m/s. Wind velocity and water-level fluctuations were measured at a fetch of x = 6.5 m from the edge (x = 0 m) of the entrance slope plate.

A laser Doppler anemometer (Dantec LDA) and phase Doppler anemometer (Dantec PDA) were used to measure the wind-velocity fluctuations. And the shear stress at the interface was estimated by an eddy correlation method [5].

Water level fluctuations were measured using resistance-type wave gauges (Kenek CHT4-HR60BNC). The resistance wire was placed in the water, and the electrical resistance due to the instantaneous water level was recorded at 500 Hz and 600 s on a digital recorder (Sony EX-UT10). The energy of the wind-waves (E) was estimated by integrating the spectrum of water level fluctuations against the frequency (f) and the significant frequency of wind-waves (f_m) was defined as the peak frequency of the spectrum.

To determine whether the dispersed droplets and entrained bubbles from the intensively breaking wind-waves caused problems for the wave gauge, we photographed the surface of breaking wind-waves with a high-speed camera (DANTEC NanoSense Mk3) synchronized to the gauge. The gauge was mounted at 0.02 m from the glass side wall of the wind-wave tank. The high-speed camera was operated at a speed of 100 frames per second to capture the surface of breaking wind-waves at the side wall of the tank. The sampling time was 20 s. The spatial resolution of a digital CMOS camera was 1280 × 1024 pixels for the imaging area of 130 mm × 100 mm (U_{10} < 30 m/s) and 260 mm × 200 mm ($U_{10} \gtrsim$ 30 m/s). The instantaneous surface elevation from the still water level was estimated from the image by eye, because it was difficult to calculate it using an analytical program owing to the intensive breaking of wind-waves.

For measuring both the wave length (L_S) and phase velocity (C_P) of significant wind-waves, another wave gauge was fixed downstream at Δx = 0.02 m for U_{10} < 30 m/s and at Δx = 0.19 m for $U_{10} \gtrsim$ 30 m/s. The values of C_P were estimated using the cospectrum method [e.g., 16].

2.2. *Field experiment*

We analysed wind and wave data recorded on an observation tower operated by the Shirahama Oceanographic Observatory, Disaster Prevention Research Institute (DPRI), Kyoto University, Japan. The tower is located at 33.4232°N and 135.1958°E in Tanabe Bay, Wakayama, Japan. The distance from the tower to the nearest coastline is about 1 km and the water depth at the tower is 30 m. The data period was from 1 January 2005 to 31 December 2009. A sonic anemometer was installed on the tower at a height of 20 m above the mean sea surface. The two horizontal and one vertical components of wind velocity were recorded at a sampling frequency of 10 Hz. The U_{10}, u^*, and wind-wave spectrum were calculated according to the methods described in Takagaki *et al.* [5].

3. Results and discussion

Figure 1 shows the wind-wave spectra at normal (U_{10} = 17 m/s) and extremely high (U_{10} = 48, 67 m/s) wind speeds obtained by wave-gauge measurements. The spectra and frequencies are normalized by the total energy (E) and significant frequency (f_m) of the wind-waves. Three solid lines showing slopes of –3, –4, and –5 at frequencies higher than f_m are described in figure 1. At normal wind speeds, the present slope at frequencies higher than f_m corresponds to the –4 slope instead of the –3 and –5 slopes (figure 1a). The –5 slope, proposed by Phillips [17], has been used in many models for wind-wave spectra [e.g., 18, 19]. The –4 slope, proposed by Toba [20] in relation to Toba's 3/2 law [21], indicates the local equilibrium between wind shear and wind-waves. It has been shown to be suitable for representing the wind-wave spectra for oceans and wind-wave tanks at normal wind speeds of $U_{10} < 35$ m/s [e.g., 22–25]. However, at extremely high wind speeds of $U_{10} \gtrsim 35$ m/s, the magnitude of the slope at frequencies higher than f_m is unknown. In figure 1b and 1c, the slope at extremely high wind speeds corresponds to the –4 slope within the region immediately after the spectral peak. This means that in the region immediately after spectral peak, a spectral energy balance mechanism [e.g., 26], similar to that at normal wind speeds, might also operate at extremely high wind speeds. On the other hand, the spectral slope becomes gentler in the region at $f > 5$ Hz. Although the gentler slope at $f > 5$ Hz may be due to measurement error or enhancement of the energy of wind-waves by the particular breaking of significant wind-waves, we cannot investigate the phenomenon here because of a lack of information on the development of wind-waves at extremely high wind speeds. Thus, the spectra at $f > 5$ Hz were eliminated from figure 1b and 1c.

Generally, the wind-wave spectral form at a slope of –4 (see figure 1) is shows as:

$$S(f) = \alpha_T (2\pi)^{-3} g u^* f^{-4} \tag{2}$$

or

$$S(f) = \alpha_D (2\pi)^{-4} g^2 f_m^{-1} f^{-4}, \tag{3}$$

where g is the acceleration due to gravity and u^* is air-friction velocity. In addition, equilibrium range constants α_T and α_D are the so-called Toba's constant and Donelan's constant, respectively. Although Toba's constant α_T was assumed to be a constant value [e.g., 22], many recent studies suggest that its value depends on inverse wave age (U_{10}/C_P) [e.g., 12, 25, and 27]. In addition, Donelan's form shown as equation (3) is suitable for estimating U_{10} (or u^*) from the wind-wave spectrum, because the form

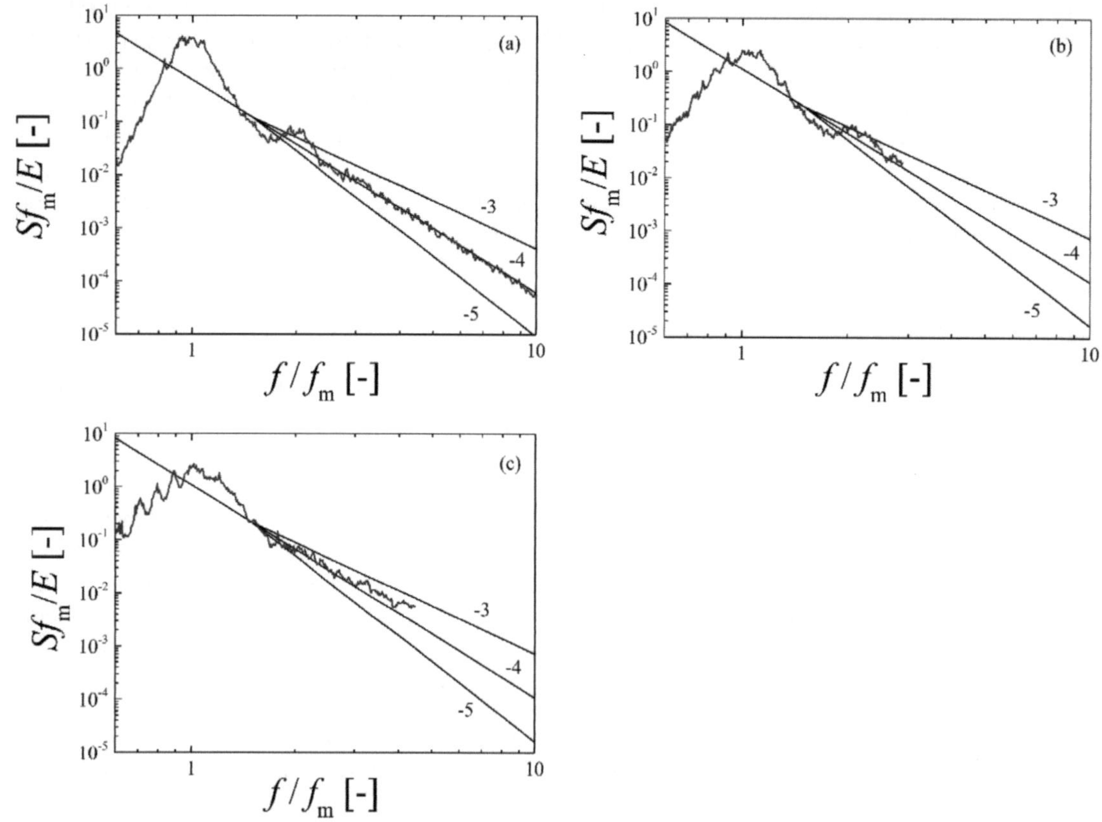

Figure 1. Wind-wave spectra at (a) normal wind speeds (U_{10} = 17 m/s), (b) extremely high wind speeds (U_{10} = 48 m/s) and (c) extremely high wind speeds (U_{10} = 67 m/s). Spectra (S) and frequency (f) are normalized by the total energy (E) and significant frequency (f_m) of wind-waves. Spectra at extremely high wind speeds (U_{10} = 48, 67 m/s) are described under f = 5 Hz by rejecting impingement effects of droplets and bubbles on water-level measurements. Three solid lines (upper, middle, and lower) are the slopes of –3, –4, and –5.

does not include a wind speed parameter directly. Therefore, in the present study, we chose Donelan's form (equation 3) and Donelan's constant α_D to develop a model of the equilibrium range constant and propose an estimation method for U_{10} (or $u*$).

Figure 2 shows the relationship between the dimensionless equilibrium range constant (αD), showing energy levels within the equilibrium range, and inverse wave age (U10/CP). The values of αD were estimated from the wind-wave spectra in the equilibrium range [e.g., 12, 25, and 27]. The present laboratory/field data at normal wind speeds of U_{10} < 35 m/s increase monotonically with increasing U_{10}/C_P, although the present field data are somewhat scattered and have lesser values than previous field studies [23 – 25, 28, and 29]. On the other hand, present laboratory data at extremely high wind speeds of $U_{10} \gtrsim 35$ m/s are plotted on a smaller range of U_{10}/C_P, and take constant values of both α_D and U_{10}/C_P, ($\alpha_D \sim 0.05$ at $U_{10}/C_P \sim 30$). This trend of taking constant values is one of the reasons for the collapse of the local equilibrium of wind and wind-waves [5] at extremely high wind speeds. Then, from this trend of α_D at extremely high wind speeds, we can produce the best-fit curve:

$$\alpha_D = 0.00136 \left(\frac{U_{10}}{C_p}\right)^{1.0} \quad (5)$$

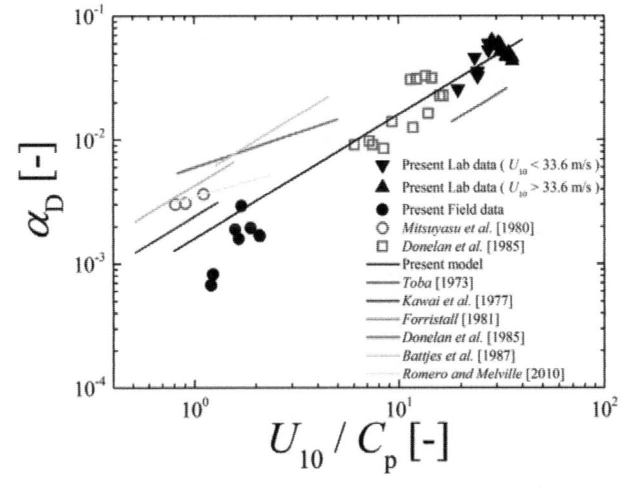

Figure 2. Relationship between the dimensionless equilibrium range constant (α_D), showing energy levels within the equilibrium range, and inverse wave age (U_{10}/C_P). The values of α_D are calculated from α_T [20, 24, 25, and 28–30] by $\alpha_D = \alpha_T\ (U_{10}/C_P)\ C_D^{1/2}$, derived from equations (2), (3), and the assumption of deep-sea waves, under the constant drag coefficient (C_D) of 0.0015. Values of α_D [30] at $U_{10}/C_P > 0.73$ are accepted here because of the limitation of the fully-developed wind-waves proposed by [31]. The solid black curve shows the best-fit against the values of α_D in present and previous [23, 30] laboratory/field studies. For Donelan *et al.* [23], the laboratory and field data are shown as open square plots and pink curve, respectively.

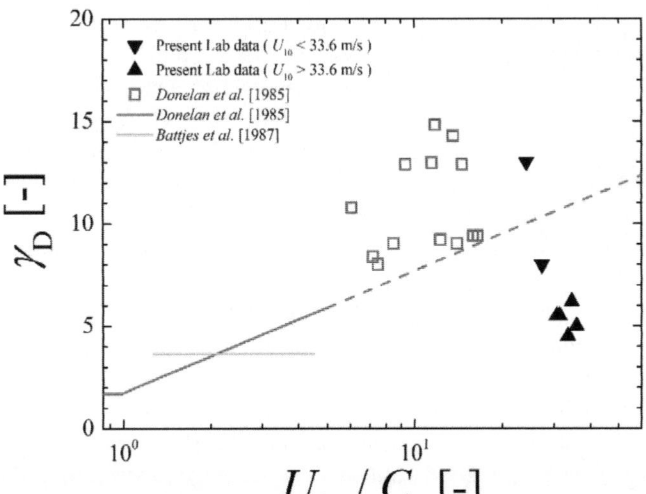

Figure 3. Relationship between the dimensionless peak enhancement factor (γ_D) and inverse wave age (U_{10}/C_P). The γ_D is evaluated according to $S(f) = \alpha_D(2\pi)^{-4}g^2 f_m^{-1} f^{-4}\exp\left(-\left(\frac{f}{fm}\right)^4\right)\gamma_D^{\Gamma}$ proposed by Donelan *et al.* [23]. The dotted curve shows the extrapolated curve of Donelan *et al.* [23]. For Donelan *et al.* [23], the laboratory and field data are shown as open square plots and pink curve, respectively.

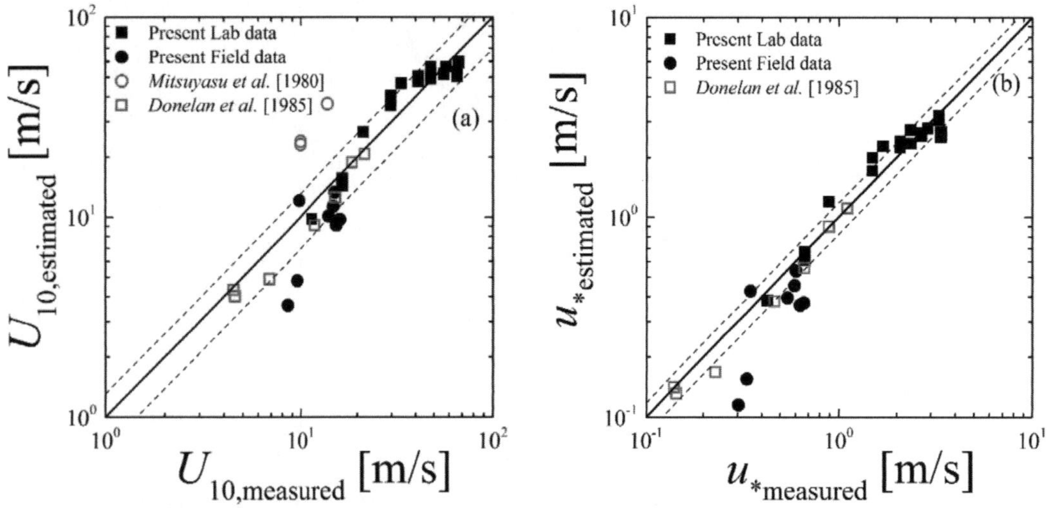

Figure 4. Relationship between measured and estimated values of (a) 10 m-height wind speed (U_{10}) and (b) air friction velocity. Data [30] are shown only in (a), because the article provides values of U_{10}. Upper and lower dashed lines in figure 4(a) and (b) show the upper and lower limitations estimated from the mean absolute percent error (30% and 18%, respectively).

from the present laboratory/field data, previous laboratory data [23], and previous field data [30] over a wide range of $U_{10}/C_P = 0.83 \sim 36$ and $U_{10} = 4.5 \sim 68$ m/s (see black curve in figure 2). The curve of equation (5) seems to be a good fit for data at normal wind speeds as well as data at extremely high wind speeds, with a correlation coefficient of $R^2 = 0.9164$. In addition, because the maximum wavelength of significant wind-waves (L_S) in the wind-wave tank is about 1.9 m at the highest wind speed (U_{10} = 68 m/s), the ratio of the water depth (D = 0.8 m) to wavelength (D/L_S) is nearly 0.5, which is a strict criterion for deep-sea waves [see 32]. This means that the wind-waves in our wind-wave tank could be regarded as almost deep-sea waves under all laboratory conditions in the present study, and that the relationship between α_D and U_{10}/C_P (figure 2) is not affected by water depth. Therefore, when α_D is measured along with C_P under tropical cyclones in real oceans, it might be possible to estimate U_{10} using equation (5). Figure 3 shows the relationship between the peak enhancement factor (γ_D) and inverse wave age (U_{10}/C_p). Although at normal wind speeds of $U_{10} < 35$ m/s, the present laboratory data increase monotonically with increasing U_{10}/C_p, present laboratory data at extremely high wind speeds of $U_{10} \gtrsim$ 35 m/s decrease rapidly with increasing U_{10}/C_p. This reduction of the peak enhancement factor at extremely high wind speeds indicates that the large-scale geometry corresponding to the significant waves is broken by intensive wind shear, and the reduction is the main reason for the collapse of the local equilibrium of wind and wind-waves [5].

Figure 4 shows the relationship between measured and estimated values of 10-m height wind speed (U_{10}) and air friction velocity (u*). Estimated values of U_{10} were calculated from the values of α_D and C_P in equation (5). In addition, estimated values of u* were calculated from estimated U_{10} and an empirical relation [5] shown as:

$$E^* = c f_m^{*-3}, \tag{6}$$

where E^* ($= E/z_0^2$) is the wave energy E normalized by the roughness length z_0, f_m^* ($=f_m(z_0/g)^{1/2}$) is the normalized significant frequency of wind-waves, $c = 1.2\times10^{-3}$, and Equation (6) means that z_0 and C_D are uniquely determined by the shape of wind-waves at both normal and extremely high wind speeds.

The correlation coefficients are $R^2 = 0.924$ for U_{10} and $R^2 = 0.966$ for u^*, and this means that the values of U_{10} and u^* are fairly accurate despite their estimation from wave measurements alone, without any wind measurements, at extremely high wind speeds in oceans under tropical cyclones.

In summary, by wave measurements in oceans under tropical cyclones, U_{10} can be estimated as:

$$U_{10} = \frac{\alpha_D}{0.00136} \frac{g}{2\pi f_m}, \quad (7)$$

which is derived from equation (5), the dispersion relation for deep-sea waves ($2\pi f_m{}^2 = g/L_S$), and $C_P = f_m L_S$. The friction velocity (u^*) can be estimated as:

$$u^* = \kappa U_{10} \left(\ln \frac{10}{z_0}\right)^{-1} = \kappa \left(\frac{\alpha_D}{0.00136} \frac{g}{2\pi f_m}\right) \left(\ln \frac{10}{c^{-2} E^2 f_m^6 g^{-3}}\right)^{-1}, \quad (8)$$

where κ (=0.4) is the von Kármán constant, and equation (8) is derived from equations (6) and (7) and the log-law under neutral conditions. Equations (7) and (8) show that U_{10} and u^* can be estimated from wave measurements alone, because α_D can be estimated by equation (3) and E can also be estimated by integrating the spectrum of water level fluctuations. We expect that equations (7) and (8) will support fundamental ocean studies and improve the precision of weather forecasting under tropical cyclones.

Acknowledgments
This work was supported by the Ministry of Education, Science, Sports and Culture, Grant-in Aid (25249013, 24360069). We gratefully acknowledge Y. Muto, Y. Baba, S. Serizawa and T. Kubo of Shirahama Oceanographic Observatory, Disaster Prevention Research Institute, Kyoto University, for providing us with the data observed at Tanabe-Nakashima oceanographic observation tower.

References
[1] Powell M D *et al* 2003 *Nature* **422** 279
[2] Hawkins H F and Rubsam D T 1968 *Mon. Wea. Rev.* **96** 617
[3] Kunishi H and Imasato N 1966 *Disaster Prevention Res. Inst. Ann.* **9** 1
[4] Donelan M A *et al* 2004 *Geophys. Res. Lett.* **31** doi:10.1029/2004GL019460
[5] Takagaki N *et al* 2012 *Geophys. Res. Lett.* **39** doi:10.1029/2012GL053988
[6] Komori S *et al* 1993 *J. Fluid Mech.* **249** 161
[7] Takagaki N and Komori S 2007 *J. Geophys. Res.* **112** doi:10.1029/2006JC003752
[8] Holthuijsen L H *et al* 2012 *J. Geophys. Res.* **117** doi:10.1029/2012JC007983
[9] Richter D H and Stern D P 2014 *Geophys. Res. Lett.* **41** doi: 10.1002/2014GL059746.
[10] Young I R 1993 *J. Geophys. Res.* **98** 20275
[11] Zhao D and Toba Y 2003 *J. Oceanogr. Soc. Japan* **59** 235
[12] García-Nava H *et al* 2012 *J. Geophys. Res.* **117** doi:10.1029/2011JC007833
[13] Thomson J *et al* 2013 *J. Geophys. Res.* **118** doi:10.1002/2013JC008837
[14] Iwano K *et al* 2013 *Tellus* B **65** 21341
[15] Krall K E and Jähne B 2014 *Ocean Sci.* **10** 257
[16] Mitsuyasu H *et al* 1979 *J. Fluid Mech.* **92** 731
[17] Phillips M 1958 *J. Fluid Mech.* **4** 426
[18] Pierson W J and Moskowitz L 1964 *J. Geophys. Res.* **69** 5181
[19] Hasselmann K *et al* 1973 *Erganzugsh. Dtsch. Hydrogr. Z.* **12** 1-95
[20] Toba Y 1973 *J. Oceanogr. Soc. Japan* **29** 209
[21] Toba Y 1972 *J. Oceanogr. Soc. Japan* **28** 109
[22] Phillips O M 1985 *J. Fluid Mech.* **156** 505
[23] Donelan M A *et al* 1985 *Philos. Trans. R. Soc. London* A **315** 509

[24] Battjes J A *et al* 1987 *J. Phys. Oceanogr.* **17** 1288
[25] Romero L and Melville W K 2010 *J. Phys. Oceanogr.* **40** 441
[26] Holthuijsen L H 2007 *Waves in Oceanic and Coastal Waters* (Cambridge: Cambridge Univ. Press)
[27] Resio D T *et al* 2004 *J. Geophys. Res.* **109** doi:10.1029/2003JC001788
[28] Kawai S *et al* 1977 *J. Oceanogr. Soc. Japan* **33** 137
[29] Forristall G Z 1981 *J. Geophys. Res.* **86** 8075
[30] Mitsuyasu H *et al* 1980 *J. Phys. Oceanogr.* **10** 286
[31] Wilson B W 1965 *Dt. Hydrogr. Z.* **18** 114
[32] Lamb H 1932 *Hydrodynamics* 6th ed. (Cambridge: Cambridge University Press)

IOP Conf. Series: Earth and Environmental Science **35** (2016) 012010 doi:10.1088/1755-1315/35/1/012010

Air-sea gas transfer for two gases of different solubility (CO_2 and O_2)

A Rutgersson[1], A Andersson and E. Sahlée

Department of Earth Sciences, Uppsala University, Uppsala, Sweden

E-mail: anna.rutgersson@met.uu.se

Abstract. At the land-based marine measuring site Östergarnsholm in the Baltic Sea, the eddy covariance technique was used to measure air-sea fluxes of carbon dioxide and oxygen. High-frequency measurements of oxygen were taken with a Microx TX3 optode using the luminescence lifetime technique. The system gives reasonable oxygen fluxes after the limited frequency response of the sensor was corrected for. For fluxes of carbon dioxide the LICOR-7500 instrument was used. Using flux data to estimate transfer velocities indicates higher transfer velocity for oxygen compared to carbon dioxide for winds above 5 m/s. There are too few data for any extensive conclusions, but a least-square fit of the data gives a cubic wind speed dependence of oxygen corresponding to $k_{660} = 0.074U_{10}^3$. The more effective transfer for oxygen compared to carbon dioxide above 5 m/s is most likely due to enhanced efficiency of oxygen exchange across the surface. Oxygen has lower solubility compared with carbon dioxide and might be more influenced by near surface processes such as microscale wave breaking or sea spray.

1. Introduction

Air-sea exchange of various gases is important for many applications, i.e., in atmospheric chemistry, marine biology and climate. Fluxes of carbon dioxide are the subject of continuing experimental and theoretical interest due to the link with global climate. Air-sea exchange of carbon dioxide (CO_2) has the obvious importance for atmospheric concentration and thus the radiation budget. Oxygen (O_2) in the ocean is of crucial importance as it is strongly linked to both carbon and nitrogen cycles [1,2] and a key component in eutrophication. As O_2 is always present in the atmosphere at fairly constant concentrations, the atmosphere acts as a sink or source depending mainly on the processes in the water. Results from more than a decade of direct measurements using the eddy covariance technique (EC) of CO_2 in marine environments show that there are still many challenges needing to be addressed, both in terms of methodology and physical understanding of the exchange processes. Most marine EC data are taken from ships and Blomquist *et al.* [3] demonstrates experimental approaches for improving measurement precision and accuracy. Several of the problems are, however, avoided when taking the measurements on a land based stationary tower. In contrast to CO_2 there are very few existing studies using EC measurements for O_2, as the technique for applying fast response sensors is very new [4].

During recent decades much effort has gone into determining the relationship between the air-sea gas transfer velocity and different dynamical forcing processes, see e.g., Garbe *et al.* [5]. Gas transfer velocity describes the efficiency of the gas transfer across the interface. Over the years numerous laboratory and field studies have been conducted and several expressions suggesting relating the gas transfer to different physical variables, such as sea spray and bubbles [6, 7, 8, 9], surface films [10, 11],

Content from this work may be used under the terms of the Creative Commons Attribution 3.0 licence. Any further distribution of this work must maintain attribution to the author(s) and the title of the work, journal citation and DOI.
Published under licence by IOP Publishing Ltd

water-side convection [12, 13] and rain [14]. Still wind speed is the most robust single parameter to describe the transfer velocity. The relative importance of these processes for air-sea exchange is likely to vary for gases of different solubility and is still not fully understood [5]. It is well known that water vapour and heat (corresponding to very high solubility) have less wind speed dependence than CO_2 (e.g., [15]), acetone with high solubility [16] has a smaller wind speed dependence. O_2 has a lower solubility than CO_2, but is frequently assumed to follow the same wind speed dependence. Previous estimates of the transfer velocity of O_2 are determined using a budget method [17] or eddy covariance fluxes in the water [18].

We here use data taken using the EC method of carbon dioxide and oxygen at the land based site Östergarnsholm. By simultaneously using gases of different solubility we expect to expand knowledge on the important processes for transfer velocities of gases in general. Few previous studies exist using EC measurements of oxygen in the atmosphere. Andersson *et al.* [4], however, show the potential of using a high-frequency probe, based on a luminescence lifetime technique.

2. Methods and data

2.1. Theory

Air-sea fluxes of gases (F) are typically calculated from concentration gradients in either the air side or water side using the two-layer model [19]:

$$F = K_a\left(\frac{C_w}{H} - C_a\right) = K_w(C_w - HC_a) \quad (1)$$

C_w and C_a are the gas concentrations in bulk water and air, H is the dimensionless Henrys solubility. The total gas transfer velocity from the perspective of the air-side gradient (K_a) or water-side concentration gradient (K_w) are related by $K_a = HK_w$. The total gas transfer velocities are composed of the individual transfer velocities in the air phase and water phase (k_a and k_w, respectively).

$$K_a = \frac{1}{(1/k_a + 1/Hk_w)}; \quad K_w = \frac{1}{(1/k_w + 1/Hk_a)} \quad (2)$$

CO_2 and O_2 are both gases with low solubility and thus water-side controlled (i.e., $K_w \approx k_w$), we here denote the total gas transfer velocity k and assume it to be equal to the water-side transfer coefficient of the gas. Rewriting equation 1 in terms of the gas partial pressure we get

$$k = \frac{F}{K_0 \Delta pC} \quad (3)$$

where K_0 is the gas specific solubility constant and ΔpC the air-sea difference of the partial pressure.

2.2. Site

The measurements used in this study are taken at the Östergarnsholm site in the Baltic Sea. This station is located at 57°27'N, 18°59'E (see figure 1). The site has been running semi-continuously since 1995; it is a land-based 30-m tower situated on the southern tip of a very small, flat island in the Baltic Sea. Flux of a quantity measured at a particular height is influenced by the surface conditions at a certain distance upwind; this area of influence is called the flux footprint. The relative role of upwind areas (flux footprint) can be estimated by applying expressions originally developed for atmospheric dispersion. For wind directions 80°<WD<210°, the data have been shown to represent open sea conditions, in the sense that the wave field is mainly undisturbed and the atmospheric turbulence and the fluxes of momentum, sensible heat, and latent heat are not influenced by either the limited water depth or the coast [21]. In Rutgersson *et al.* [20] CO_2 measurements were investigated, showing that for wind from the 80°<WD<160° sector, the sensor for pCO_{2w} (partial pressure of CO_2 in the water) represents the footprint of the tower. The present investigation uses only data for winds from the 80°<WD<160° sector.

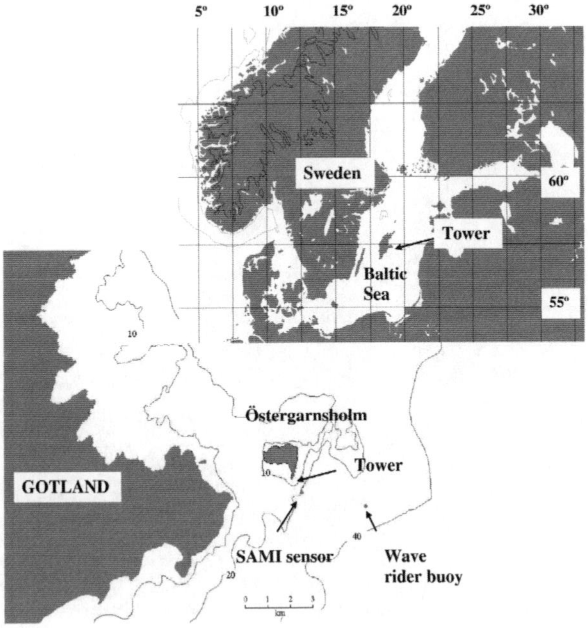

Figure 1. Map of the Östergarnsholm measurements site (from [20]). Positions of the tower and SAMI sensor are indicated by arrows. Thin solid lines represents iso-lines of water depth.

The tower is equipped with high-frequency instrumentation for turbulence measurements and slow-response sensors for mean profile measurements. High-frequency wind components are measured with CSAT3 (Campbell Scientific, North Logan, Utah, USA) sonic anemometers at three levels, i.e., 9, 16.5, and 25 m above the tower base, for earlier data the SOLENT 1012R2 sonic anemometers (Gill Instruments, Lymington, UK) were used. Sea level varies slightly, but the tower base is generally 1±0.5 m above mean sea level. The site and instrumentation protocols are further described in Rutgersson *et al.* [20]. The sampling rate of the eddy covariance system was set to 20 Hz and fluxes were calculated over 60-min intervals. There was careful quality control for all individual flux data points.

2.3. CO_2 data

The humidity and CO_2 fluctuations are measured with a LICOR-7500 (LICOR-Inc., Lincoln, NE, USA) open-path analyzer at 9 m above the tower base. For pCO_{2w} and sea surface temperature (SST) a SAMI sensor (Sunburst Sensors, Missoula, USA) is used [22]. It measures at a depth of approximately 4 m, 1 km SE of the tower (see figure 1). The investigated area is a sink or source of atmospheric CO_2 depending on the season, pCO_{2w} displays a significant seasonal cycle with summertime values of 150 μatm and up to 800 μatm in winter. A smaller seasonal cycle is observed in the atmosphere, with a peak-to-peak amplitude of 17 μatm [23]. The air-sea difference in partial pressure is relatively large (of the order of 50–200 μatm).

2.4. O_2 data

The high-frequency oxygen data were measured at the height of 25 m with the commercially available optode system Microx TX3 (PreSens- Precision Sensing GmbH, Regensburg, Germany). The Microx TX3 sampling rate is 2 Hz with a precision of 0.1% air saturation. The sensor was, however, found to have a limited response time, yielding a flux loss in the frequency range f> 0.3 Hz. This is corrected for by applying cospectral similarity simultaneously using measurements of latent heat as the reference signal. On average the magnitude of the cospectral correction added 20% to the uncorrected oxygen flux (see [4] for details of the system and corrections). The measured fluxes were also density corrected according to Webb *et al.* [24], a significant term, sometimes also changing the direction of the fluxes. Figure 2 shows the set-up of the system with the CSAT3 sonic (figure 2a), the LICOR at the same height (figure 2b), and the needle of the oxygen sensor (figure 2c).

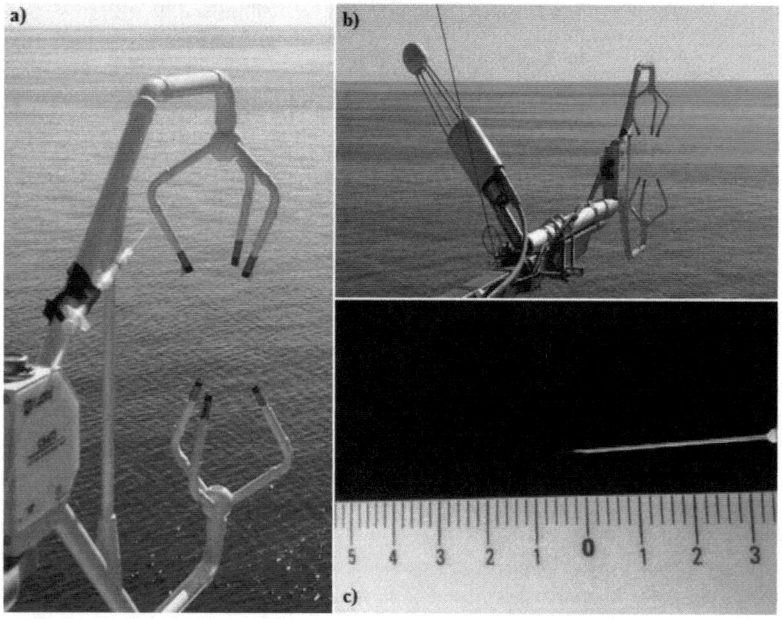

Figure 2. (a) The EC system with a close-up of the oxygen sensor mounted at the CSAT3, (b) the EC system at the height of 25 m, and (c) a close-up of the oxygen sensor tip. Figure from [4].

The water-side oxygen is measured by a SBE-43 sensor integrated on a SBE 37 smp-ido (Seabird Electronics Inc., Bellevue, Washington, USA) at 4 m depth placed at the mooring of the SAMI.

3. Results

3.1. Case study

During a field campaign in June 2013 fluxes of CO_2 and O_2 were measured simultaneously. The campaign was from June 19 to June 24 and the conditions are shown in figure 3. The data are characterized by three periods, the second period, 21 June to 23 June (decimal month 6.69 to 6.74), represents an upwelling period. The wind increases and the direction changes to 200° for more than 30 hours, this is typical for upwelling situations in this region [25]. The upwelling is also clearly seen in satellite SST maps for the period (not shown). As the buoy measurements are not representative for the flux footprint of the tower during the upwelling situation, these data are not used in the analysis of transfer velocity and the focus is on the first and third periods. The first period shows positive (upward directed) oxygen fluxes of 0.6–6.4 μmol $m^{-2}\ s^{-1}$, in agreement with the oxygen saturation in water varying between 105–108% (in figure 3 oxygen partial pressure is presented). After the upwelling period, oxygen saturation increases from 5% supersaturation to leveling off at 10% supersaturation when the water temperature goes back to 14°C during the third period and the oxygen fluxes are 0.35–-3 μmol $m^{-2}\ s^{-1}$. The partial pressure of CO_2 is significantly lower in the water compared to air, $\Delta pCO_2 > 150$ μatm, indicating large fluxes and a large signal-to-noise ratio of the data.

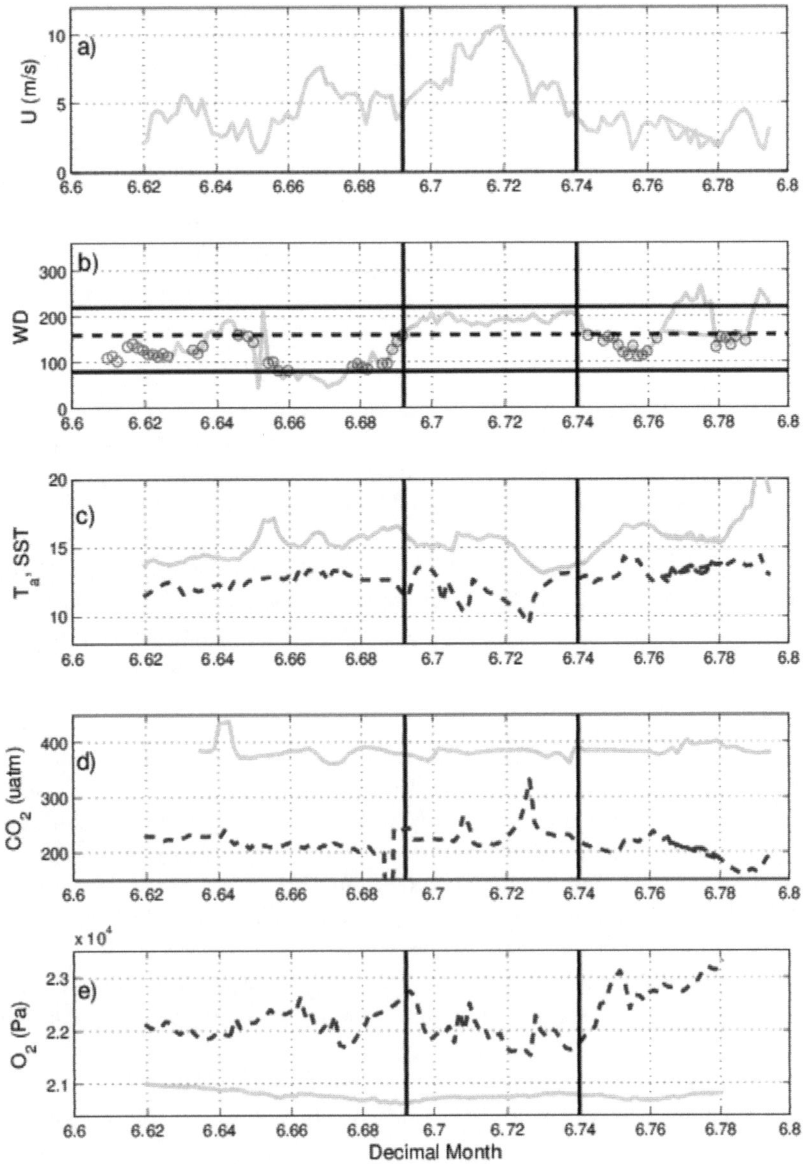

Figure 3. Field campaign June 19 to 24, 2013. Wind speed (a), wind direction (b), temperature (c), partial pressure of CO_2 (d) and O_2 (e). Green lines correspond to data taken in the tower and blue (dashed) curves to measurements in the water. Vertical black lines represent an upwelling period and horizontal black lines in (b) indicate the wind directions where measurements in the water are representative for conditions at the tower (these data are also shown by red circles).

Using the fluxes measured with the EC method and the air-sea differences of partial pressure the transfer velocities can be calculated according to equation 3. The transfer velocities of CO_2 and O_2 (denoted k_{660} when normalized to a Schmidt number of 660) are shown for various wind speeds in figure 4a. Shown also is the wind speed dependent equation $k_{660} = 0.26U_{10}^2$ derived for CO_2, where U_{10} is the wind speed at 10 m.

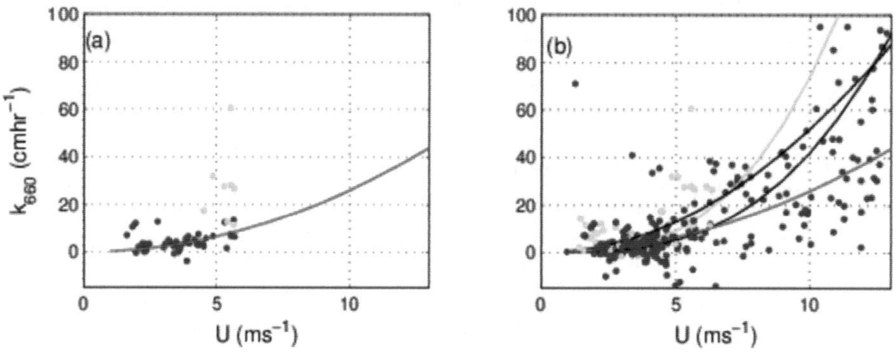

Figure 4. Transfer velocity of CO_2 (blue dots) and O_2 (green dots) for varying wind speeds for the field campaign (a) and for the extended data set (b). Red lines represent the wind speed dependent equation $k_{660} = 0.26U_{10}^2$. In (b) and green curve is $k_{660} = 0.074U_{10}^3$ and black lines represent expressions for O_2 from [17], corresponding to cubic (dashed) and quadratic (solid) equations, respectively.

There are few data for the oxygen transfer velocity and significant scatter, but the significantly higher values of k_{660} for O_2 compared to k_{660} for CO_2 for the same wind speeds indicate a more efficient transfer across the surface for the gas with the lower solubility. Both gases here experience the same environmental conditions because the measurements were taken simultaneously.

3.2. Extended data set

For a more extended analysis, several data sets are used. For CO_2 this includes data from 2005–2007 described in [12] as well as more recent data from 2013–2014, for O_2 this includes data from July 2011 and September 2011 in addition to June 2013. All data are taken at the Östergarnsholm site. Situations with water-side convection (defined as in [13]) have been removed as these data represent different conditions and no data exist for O_2 during such conditions. Figure 4b shows the transfer velocities of CO_2 and O_2 and the extended data set supports the case study. The data of O_2 are too few to draw any final conclusion, but figure 4 indicates a stronger wind speed dependence of O_2 than CO_2. A fit to the O_2 data points indicates a cubic wind speed dependence $k_{660} = 0.074U_{10}^3$ (shown by the green line in figure 4b), the O_2 transfer velocity is larger than for CO_2 at winds approximately larger than 5 m/s. Also shown are the two curves derived for O_2 in [17] using a cubic and quadratic wind speed dependence ($k_{660} = 0.042U_{10}^3$ and $k_{660} = 0.52U_{10}^2$, respectively). For CO_2 the scatter increases with increasing winds and for winds above 8 m/s some of the data points are well above the expected quadratic wind speed dependence (the red curve).

4. Discussion

For winds within the regime 5–7 m s^{-1} the measured transfer velocity for O_2 shows a stronger wind speed dependence than for CO_2 and also stronger compared to previous parameterizations developed for O_2. The resistance to transfer of gases of relatively low solubility across the sea surface can be seen as resistances in an electric circuit (e.g., [13]) and different processes are of different relative importance depending of the solubility of the gas. It is likely that processes enhancing the exchange more effectively for gases with a lower solubility starts to be of importance already at these wind speeds. Examples of such processes that could be active for our data are microscale wave breaking, bubbles or sea spray. The onset of breaking waves and whitecap coverage are generally found at lower wind speeds for coastal regions and marginal seas, which could explain the strong wind dependence at 5 m s^{-1}. The short averaging time scale in the EC method (here 60 min) makes it possible to relate the measured transfer velocity to episodic high wind events and to study the impact of processes acting on higher temporal resolution.

In Kihm and Körtzinger [17], the importance of high wind speeds for air-sea gas exchange of O_2 was pointed out when suggesting their cubic relation (giving significantly lower transfer velocities compared to our measurements). It is, however, likely that the effect from enhanced mixing with short time scales on the oxygen transfer velocity in [17] were underestimated as their results are based on a budget method with a much longer averaging time scale (7 days). Also for CO_2 the scatter at winds above 8 m/s can probably be explained by the presence of processes enhancing the efficiency of the mixing for some of the data.

5. Conclusions

High-frequency measurements of oxygen were taken with a Microx TX3 optode using a luminescence lifetime technique; the system gives reasonable oxygen fluxes after correcting for the limited frequency response. Using flux data to estimate the transfer velocity for oxygen indicates a higher transfer velocity for O_2 compared to CO_2 for winds above 5 m/s. There are too few data for extensive conclusions, but a least-square fit of the data gives a cubic wind speed dependence corresponding to $k_{660} = 0.074U_{10}^3$. The more effective transfer for O_2 compared to CO_2 above 5 m/s is most likely due to enhanced efficiency of air-sea transfer of O_2 with lower solubility due to microscale breaking or sea spray.

References

[1] Arrigo K R 2004 *Nature* **437** 349
[2] Keeling R F *et al* 2010 *Mar. Sci.* **2**
[3] Blomquist B W *et al* 2014 *Bound. Layer Meteorol.* **152** 245
[4] Andersson A *et al* 2014 *J. Atmos. Ocean. Technol.* **31** 2498
[5] Garbe C S *et al* 2014 *Ocean-Atmosphere Interactions of Gases and Particles* eds P Liss and M Johnson (Springer-Verlag) 55-112
[6] Monahan E C and Spillane M C 1984 *Gas Transfer at Water Surfaces* eds W Brutsaert and G H Jirka (Spinger-Verlag) 495–503
[7] Wallace D W R and Wirick C D 1992 *Nature* **356** 694
[8] Woolf D K *et al* 2007 *J. Mar. Syst.* **66** 71
[9] Asher W E and Wanninkhof R 1998 *J. Geophys.Res.* **103** 10555
[10] Frew N M 1997 *The Sea Surface and Global Change* eds P S Liss and R A Duce (Cambridge University Press) 121–72
[11] Salter M *et al* 2011 *J. Geophys. Res.* **116** C11016
[12] Rutgersson A and Smedman A 2010 *J. Mar. Syst.* **80** 125
[13] Rutgersson A *et al* 2011 *Geophys. Res. Lett.* **38** L02602
[14] Takagaki N and Komori S 2007 *J. Geophys. Res.* **112** C06006
[15] Rutgersson A *et al* 2001 *Bound.-Layer Meteor.* **99** 53
[16] Yang M *et al* 2014 *J. Geophys. Res.* **119** 7308
[17] Kihm C and A Körtzinger 2010 *J. Geophys. Res.* **115** C12003
[18] McNeil C D and D'Asaro E 2007 *J. Mar. Syst.* **66** 110
[19] Liss P S and P G Slater 1974 *Nature* **247** 181
[20] Rutgersson A M *et al* 2008 *J. Mar. Syst.* **74** 381
[21] Högström U *et al* 2008 *Boreal Environ. Res.* **13** 475
[22] DeGrandpre M *et al* 1995 *Limnol. Oceanog.* **40** 969
[23] Rutgersson A *et al* 2009 *Bor. Environ. Res.* **14** 238
[24] Webb E K *et al* 1980 *Q. J. R. Meteorol. Soc.* **106** 85
[25] Norman M *et al* 2013 *Tellus* B **65**

IOP Conf. Series: Earth and Environmental Science **35** (2016) 012011 doi:10.1088/1755-1315/35/1/012011

Air-sea transfer of gas phase controlled compounds

M Yang[1], T G Bell[1], B W Blomquist[2], C W Fairall[2], I M Brooks[3] and P D Nightingale[1]

[1] Plymouth Marine Laboratory, Prospect Place, Plymouth, UK
[2] NOAA Earth System Research Laboratory, Boulder, Colorado, USA
[3] Institute for Climate and Atmospheric Science, School of Earth and Environment, University of Leeds, Leeds, UK

Email: miya@pml.ac.uk

Abstract. Gases in the atmosphere/ocean have solubility that spans several orders of magnitude. Resistance in the molecular sublayer on the waterside limits the air-sea exchange of sparingly soluble gases such as SF_6 and CO_2. In contrast, both aerodynamic and molecular diffusive resistances on the airside limit the exchange of highly soluble gases (as well as heat). Here we present direct measurements of air-sea methanol and acetone transfer from two open cruises: the Atlantic Meridional Transect in 2012 and the High Wind Gas Exchange Study in 2013. The transfer of the highly soluble methanol is essentially completely airside controlled, while the less soluble acetone is subject to both airside and waterside resistances. Both compounds were measured concurrently using a proton-transfer-reaction mass spectrometer, with their fluxes quantified by the eddy covariance method. Up to a wind speed of 15 m s^{-1}, observed air-sea transfer velocities of these two gases are largely consistent with the expected near linear wind speed dependence. Measured acetone transfer velocity is ~30% lower than that of methanol, which is primarily due to the lower solubility of acetone. From this difference we estimate the "zero bubble" waterside transfer velocity, which agrees fairly well with interfacial gas transfer velocities predicted by the COARE model. At wind speeds above 15 m s^{-1}, the transfer velocities of both compounds are lower than expected in the mean. Air-sea transfer of sensible heat (also airside controlled) also appears to be reduced at wind speeds over 20 m s^{-1}. During these conditions, large waves and abundant whitecaps generate large amounts of sea spray, which is predicted to alter heat transfer and could also affect the air-sea exchange of soluble trace gases. We make an order of magnitude estimate for the impacts of sea spray on air-sea methanol transfer.

1. Introduction

Many gases that exchange between the ocean and atmosphere influence our climate and air quality. In addition to carbon dioxide (CO_2) and dimethylsulfide (DMS), the oceans can be a net source or sink of very soluble organic compounds such as methanol and acetone [1], which affect the atmosphere's ability to cleanse itself of pollutants. Other soluble/reactive gases that cross the air/sea interface include sulfur dioxide (SO_2 [2]), polychlorinated biphenyls (PCBs [3]), ozone [4], and oxygenated volatile organic compounds such as formaldehyde [5], acetaldehyde [6] and glyoxal [7].

Wind blowing over the ocean provides the predominant kinetic forcing for air-sea transfer, while the thermodynamic potential for exchange is governed by the air-sea gas disequilibrium. Based on the two-

Content from this work may be used under the terms of the Creative Commons Attribution 3.0 licence. Any further distribution of this work must maintain attribution to the author(s) and the title of the work, journal citation and DOI.
Published under licence by IOP Publishing Ltd

layer model [2], the net air-sea gas flux is usually estimated from the gas transfer velocity (K) and the air-sea concentration gradient (ΔC), with a positive flux indicating sea-to-air emission:

$$Flux = K_a(C_w/H - C_a) = K_w(C_w - HC_a) \quad (1)$$

The total gas transfer velocity from the air perspective (K_a) and water perspective (K_w) are related: $K_a = HK_w$, where H is the dimensionless liquid to gas solubility. C_w and C_a are the gas concentrations in water and air. Partitioning of ΔC and resistance to transfer in the two phases depend primarily on H:

$$K_a = 1/(1/k_a + 1/(Hk_w)) \quad (2a)$$
$$K_w = 1/(1/k_w + \mathrm{H}/k_a) \quad (2b)$$

Here k_a and k_w are the individual transfer velocities in the gas phase and water phase, respectively. Exchange of sparingly soluble gases (low H) is limited by the rate of transport in water (i.e., $K_w \approx k_w$), while the exchange of highly soluble gases (high H) is limited by transport in air (i.e., $K_a \approx k_a$). CO_2 and DMS are examples of sparingly soluble (waterside controlled) gases. The transfer of DMS is thought to be a mostly interfacial process and not very sensitive to bubbles [8, 9]. CO_2 is less soluble than DMS so is subject to greater bubble-mediated exchange in addition to interfacial exchange [8]. The transfer of methanol, a gas ~500 times more soluble than DMS, is almost entirely controlled on the airside. Acetone, ~60 times more soluble than DMS, is subject to resistance both on the airside and on the waterside (~75% and 25% at 20°C, respectively).

Current understanding of airside-controlled gas transfer stems mostly from measurements of water vapor (H_2O, i.e., latent heat) and sensible heat. Similar to highly soluble gases, there is effectively no waterside resistance to heat transfer. Due in part to the much higher molecular diffusivities of gases in air than in water, turbulent resistance is relatively more important for k_a (i.e., aerodynamic resistance) than for k_w. At moderate wind speeds, the transfer velocities of heat demonstrate a near-linear relationship to wind speed and to the friction velocity (u_*). This results in fairly constant values of the dimensionless transfer coefficients for sensible heat, latent heat, and enthalpy (= sensible heat + latent heat) at around $1e^{-3}$ [10]. At wind speeds over 20 m s^{-1}, limited heat transfer measurements demonstrate a large range [11, 12]. In such high seas, model results suggest that sea spray from wave breaking plays an important role in heat transfer [13, 14]. Sea spray could also have an effect on the transfer of airside controlled compounds (analogous to the effect of bubbles upon k_w). For example, sea spray has been shown to be a source of atmospheric hydrochloric acid [15] and is thought to be a possible sink for atmospheric SO_2 [16].

Direct air-sea transfer measurements of airside controlled trace gases (i.e., not H_2O) are rare. Aircraft observations of the surface reactive SO_2 yielded an airside transfer velocity that is ~30% lower than predicted [17], illustrating an uncertainty in our understanding of k_a. Yang *et al.* (2013a) developed a novel system to measure the air-sea transfer of methanol and acetone by eddy covariance using a proton transfer reaction mass spectrometer, PTR-MS [18]. Air-sea fluxes of these compounds were quantified during the Atlantic Meridional Transect cruise (AMT-22; [1, 19]) and more recently during the High Wind Gas Exchange Study (HiWinGS; [20]). Here we present a more in-depth analysis of the HiWinGS dataset, compare results from HiWinGS to those from AMT-22, and examine the potential effects of sea spray on methanol transfer.

2. Experimental

The transfer velocity of methanol (as well as sensible heat) was determined during the AMT-22 and the HiWinGS cruises. Acetone was measured with enough precision to derive its transfer velocity only on the HiWinGS cruise. The experimental settings and methods have been described in detail previously [1, 19, 20]. Very briefly, concentrations of both compounds in the atmosphere and surface ocean were quantified by a PTR-MS. For the majority of the cruise, the PTR-MS was operated in atmospheric mode and sampling at a rate just above 2 Hz. Winds and motion were measured by a sonic anemometer (Gill

Windmaster) and Motionpak (Systron Donner) co-located with the gas inlet on the foremast of the ship. Wind velocities corrected for ship's motion ([21] followed by sequential decorrelations with the ship's motion) were used to compute the fluxes of methanol, acetone, sensible heat, and momentum by the eddy covariance method. Approximately twice a day, the PTR-MS was switched to analyze discrete water samples for dissolved concentrations of methanol and acetone [22]. Near-surface waters were taken at a few meters below the ocean surface from twice-a-day CTD casts as well as from the ship's non-toxic underway water supply. The total transfer velocities of methanol and acetone from the atmospheric perspective were computed by dividing the measured fluxes by the air-sea concentration difference following equation 1.

HiWinGS methanol and acetone data presented in this paper have been reprocessed since Yang *et al.* [20]. The major differences in this processing are 1) compute fluxes as 20-minute averages instead of hourly averages. The 20-minute averages are then binned into hourly intervals (hours with less than two valid intervals are not considered for further analysis). Signal dropout and elevated noise at high frequencies were common for the Windmaster sonic anemometer during moderate-to-heavy precipitation, which tended to coincide with very high wind speeds. The shorter averaging time afforded ~15% more useable flux data that were previously discarded due to episodic rain events; 2) correct the *w* axis of the Windmaster sonic anemometer for a calibration bias. On the advice of the manufacturer Gill (R. McKay, personal communication, 2015), we applied a bias correction to the raw *w* data of the Windmaster (+16.6% for positive *w*; 28.9% for negative *w*). The same *w* correction has also been applied to the AMT-22 dataset.

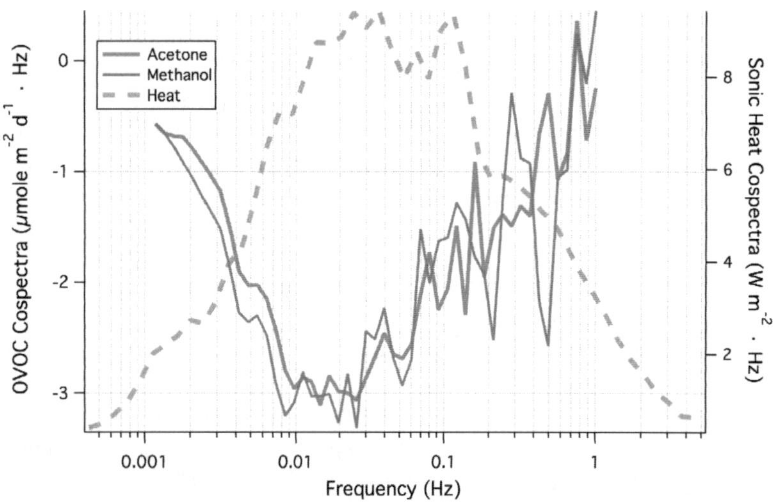

Figure 1. Mean cospectra of the oxygenated volatile organic compounds (OVOCs) methanol, acetone, and sonic heat flux during HiWinGS.

3. Results and discussion

3.1. Reprocessed HiWinGS results

The mean cospectra of methanol, acetone, and sonic heat flux for the HiWinGS cruise are shown in figure 1. Heat flux was upwards, while both methanol and acetone fluxes were downwards. The cospectra of the two gases show comparable flux magnitudes and are fairly similar in shape to that of the sonic heat flux. Due to the relatively low sampling rate of the PTR-MS, flux attenuation is evident for the gases at high frequencies, which was corrected using a filter function approach [18, 20]. For this cruise, averaging to 20 minutes instead of an hour also results in a very small loss (~1%) in the gas fluxes at low frequencies.

Compared to [20], the reprocessed methanol and acetone fluxes (as well as transfer velocities) are ~15% higher in the mean, as expected from the Gill bias correction. This correction also increases u_* by ~5%, which remains close to the predicted value from COARE 3.5 [23]. The hourly methanol and

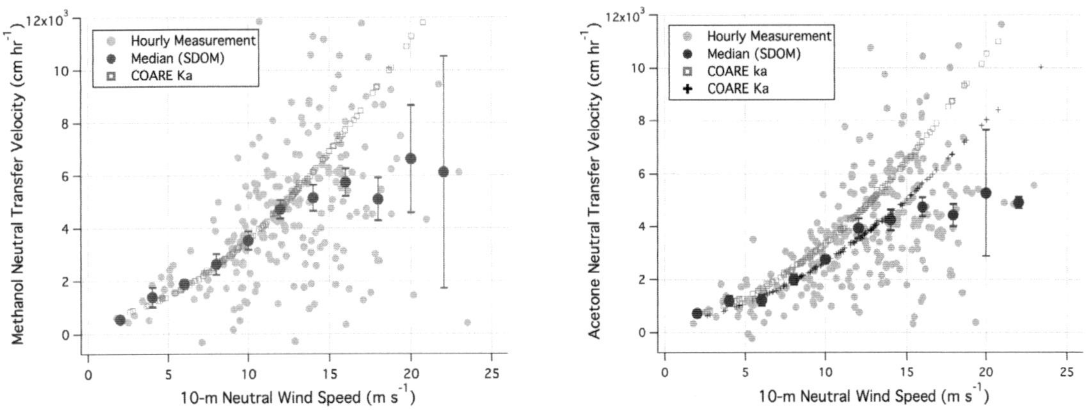

Figure 2. Methanol transfer velocity (left) and acetone transfer velocity (right) from HiWinGS.

acetone transfer velocities ($K_{METHANOL}$ and $K_{ACETONE}$) are plotted against 10-m neutral wind speed (U_{10n}) in figure 2, along with bin-medians and standard errors. They have been adjusted to a neutral atmosphere using the stability parameter from the COARE bulk output (e.g., [24]). We also plot the COARE [10] total gas transfer velocity from the atmospheric perspective (K_a) for methanol and acetone, as well as the airside transfer velocity (k_a) for acetone (note that $K_a \approx k_a$ for methanol). At wind speeds less than 15 m s^{-1}, there is close agreement between measured and predicted K_a for methanol as well as for acetone. Both show a slight non-linear dependence on wind speed (and increase essentially linearly with u_*). Between 15 and 20 m s^{-1}, measured $K_{METHANOL}$ and $K_{ACETONE}$ are lower than the model predictions, with the more soluble methanol showing a greater discrepancy. The measurement–model bias increases with wind speed. For example, the respective measurement/model ratio for $K_{METHANOL}$ is 0.74 and 0.55 at wind speeds of 16 and 18 m s^{-1}. For acetone, this ratio is 0.84 and 0.66, respectively. These results suggest a possible suppression of gas transfer that is primarily on the airside.

Gas transfer velocity data at wind speed over 20 m s^{-1} are still very limited (only six valid hours), resulting in highly uncertain bin medians. This is partly because the air-sea ΔC of these gases is dominated by their atmospheric abundance, which tended to be low in storms as a result of precipitation scavenging [20]. To reduce noise in K_a, we have neglected periods when the atmospheric mixing ratio is below 0.2 ppb. Air-to-sea (dry) deposition removes these gases from the marine atmospheric boundary layer at a timescale of 1–2 days [20]. Due to its higher solubility, wet deposition is more important of a sink for atmospheric methanol than for acetone [20]. Both of these gases have large terrestrial sources. Measuring in a region of higher atmospheric organic abundance (e.g., downwind of a continent) could help to reduce the uncertainties in $K_{METHANOL}$ and $K_{ACETONE}$.

As mentioned previously, acetone transfer is subject to significant resistance on both the air side and on the water side. From the bin-medians of $K_{ACETONE}$ and $K_{METHANOL}$, we rearrange equation 2a and compute the waterside transfer velocity $k_w = 1/(H(1/K_a - 1/k_a))$. This analysis was done previously [20] but only at a single wind speed of 12 m s^{-1} (HiWinGS mean). Here we illustrate the wind speed dependence in waterside transfer (figure 3). For this calculation we assume $K_{METHANOL} = k_a$ of acetone. We further normalize k_w to a Schmidt number of 660, e.g., $k_{w660} = k_w * (Sc_{ACETONE}/660)^{1/2}$, where $Sc_{ACETONE}$ is the ambient Schmidt number of acetone. The resultant k_{w660} should represent a "zero-bubble" (i.e., purely interfacial) waterside transfer velocity. Also shown in figure 3 are the predicted waterside transfers from the COARE gas transfer model version 3.0 (empirical constants $A = 1.3$ for interfacial transfer and $B = 1.0$ for bubble-mediated transfer) and version 3.1 ($A = 1.6$, $B = 1.8$, tangential u_* instead of total u_*) [10]. The constant B is essentially irrelevant here because bubble-mediated exchange for the very soluble acetone is effectively zero. Both versions of the COARE model fit through the HiWinGS results. Due to the large uncertainties in k_{w660} (propagated from the standard errors shown in figure 2), especially in high winds, neither version of the model performs better/worse than the other compared to observations. Uncertainties in this indirect estimation of k_{w660} could be reduced by measuring acetone in warm waters, as water side resistance becomes more important with increasing temperature.

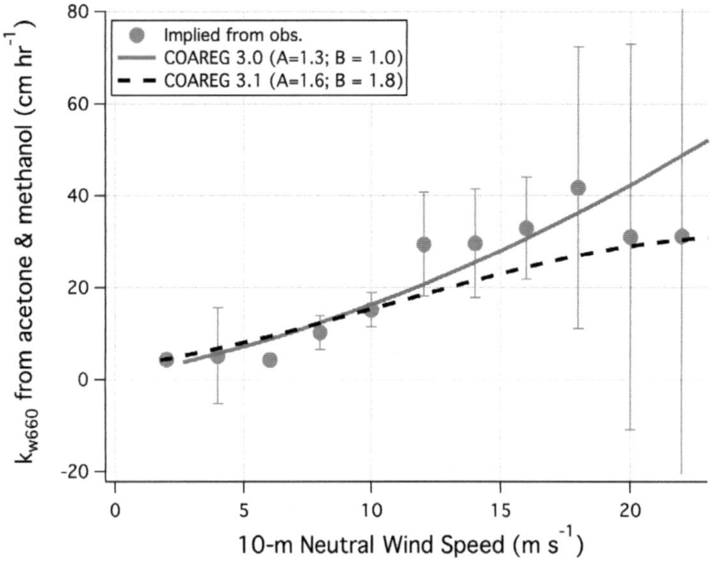

Figure 3. Indirectly estimated waterside transfer velocity from acetone and methanol transfer during HiWinGS; also shown are the modeled waterside transfer velocities of acetone at a waterside Schmidt number of 660 (COARE gas transfer model, version 3.0 and 3.1).

3.2. Comparison between HiWinGS and AMT-22

We compare HiWinGS and AMT-22 in methanol and sensible heat transfer. To reduce noise in the methanol measurement, here we compute $K_{METHANOL}$ as flux averaged over 8 hours divided by ΔC averaged over 8 hours. As in [19], for this calculation the seawater methanol concentration is set to zero for the AMT-22 cruise. At a wind speed below 15 m s^{-1}, $K_{METHANOL}$ from the two cruises demonstrate similar trends on average. The mean (±1 standard deviation) dimensionless methanol transfer coefficient ($K_{METHANOL}/U_{10n}$) is 0.98 ± 0.39 e^{-3} from AMT-22 and 1.09 ± 0.35 e^{-3} from HiWinGS.

The AMT-22 cruise only had a few hours with wind speeds over 15 m s^{-1}, during which $K_{METHANOL}$ appeared to be greater than the COARE prediction. This was initially interpreted to be due to an overestimation of the airside diffusive resistance in the COARE 3.0 model (and thus underestimation of the airside transfer velocity) [19]. The COARE model version 3.5 has a higher drag coefficient (i.e., lower aerodynamic resistance) than COARE 3.0 at high wind speeds. Implementing the COARE 3.5 drag coefficient (C_D) into the gas transfer model would also result in a slightly higher k_a to wind speed relationship than predicted using COARE 3.0. As shown in figure 4, these two model parameterizations start to diverge at a wind speed of ~15 m s^{-1}. Considering methanol observations from both cruises (with HiWinGS making up the bulk of the data at wind speeds over 10 m s^{-1}), both versions of the model appear to fit observations fairly well up to ~16 m s^{-1}.

Sensible heat transfer velocity (K_{HEAT}) was computed from the sonic heat flux (corrected for humidity using the bulk latent heat flux) and the air-sea potential temperature difference, and was further adjusted to a neutral atmosphere. As with methanol, below 15 m s^{-1} there is fairly good agreement in K_{HEAT} between the two cruises. Three parameterizations of K_{HEAT} are shown. Two are derived from airside resistance using C_D from COARE 3.0 and 3.5. The other is a product of C_{Hn} and U_{10n}, where the sensible heat transfer coefficient C_{Hn} is an empirical fit to previous open ocean transfer measurements. These parameterizations began to diverge above a wind speed of ~15 m s^{-1}, with the empirical fit noticeably lower than the resistance-based estimates. Due to the large scatter in measured K_{HEAT} at wind speeds above 15 m s^{-1}, we are unable to discern which parameterization is the most appropriate. Interestingly,

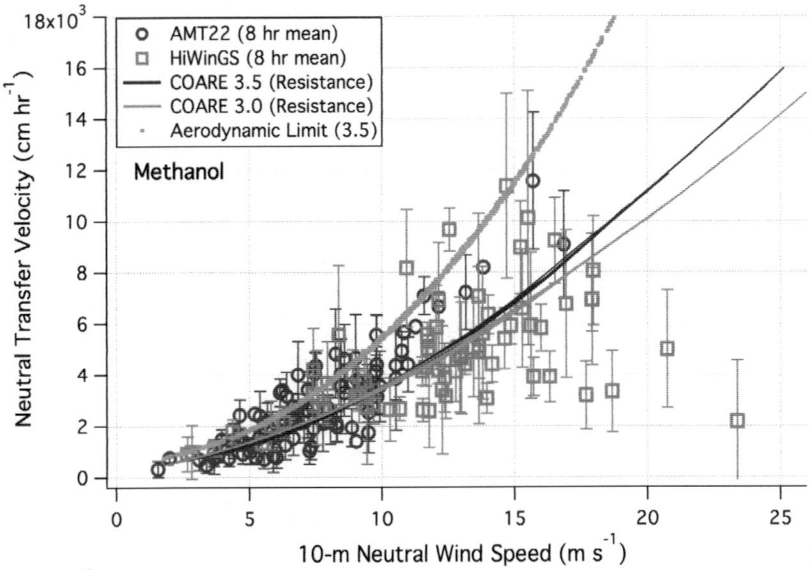

Figure 4. Transfer velocity of methanol from AMT-22 and HiWinGS cruises.

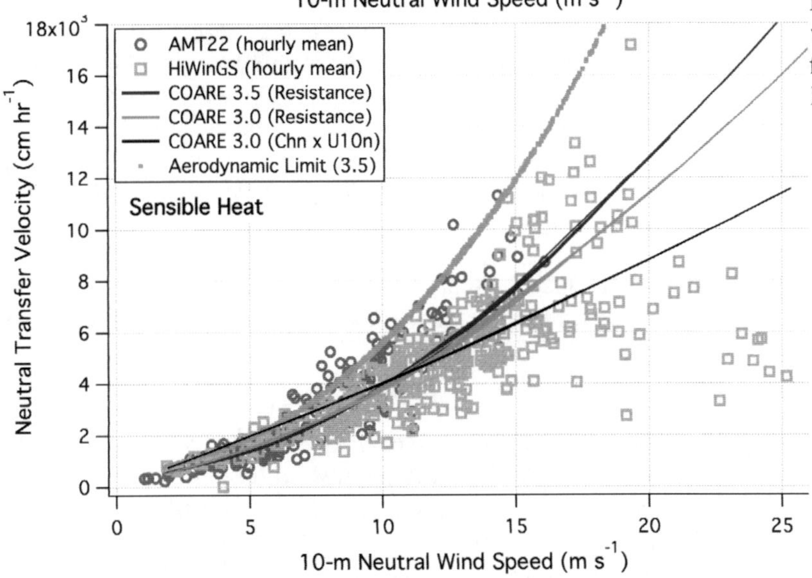

Figure 5. Transfer velocity of sensible heat from AMT-22 and HiWinGS cruises.

above a wind speed of ~20 m s^{-1}, limited K_{HEAT} measurements appear to be lower than expected from the COARE model. This is primarily because the measured sensible heat flux was ~40 W m^{-2} lower than predicted during the storm around 25 October 2013. We note that a low sensible heat transfer rate between 18 and 20 m s^{-1} has been observed in a previous study [10]. Makin (1998) also modeled a reduction in the sensible heat transfer coefficient and an increase in the latent heat transfer coefficient due to spray, which becomes important at wind speeds over 25 m s^{-1} [14]. Below we crudely examine the effects of spray on methanol transfer.

3.3. Impact of sea spray on methanol transfer

Sea spray lofted into the atmosphere is rapidly equilibrated in temperature with the surrounding air [13, 25]. When the sea surface is warmer than the air above, this leads to an initial warming of the near surface air. Partial evaporation of spray droplets gives off water vapor and cools the spray-evaporation layer (approximately equivalent to the significant wave height) over a longer timescale [25]. A net cooling in the spray-evaporation layer should reduce the vertical temperature gradient between the lowest meters of the atmosphere and the sonic anemometer (nominally ~20 m above sea level). This in theory may lead to a covariance sensible heat flux at 20 m that is lower than predicted from bulk air/water temperatures. Some methanol is likely co-emitted during spray evaporation, which could possibly reduce the vertical methanol gradient between the lowest meters of the atmosphere and the gas inlet. However, there is the competing effect of spray absorption. Droplets initially undersaturated in methanol could take up the gas from the atmosphere and enhance air-to-sea methanol deposition.

Andreas *et al.* (1995) predicted that at a wind speed of 20 m s^{-1}, the sea spray contribution to latent heat flux and sensible heat is 150 and 15 W m^{-2}, respectively [13]. The latter is of the same order of magnitude (but opposite sign) to the discrepancy of ~40 W m^{-2} in sensible heat flux (section 3.2). Dividing by the density of air and latent heat of evaporation, 150 W m^{-2} of latent heat from spray can be converted to 0.05 g of water/kg of air • m s^{-1}, or 0.06 g m^{-2} s^{-1}. Assuming spray initially carries the same dissolved methanol concentration as seawater during HiWinGS (~20 nmole L^{-1}), 0.06 g of water would contain 1.2 pmole of methanol. This implies a spray-mediated methanol emission of 1.2 pmole m^{-2} s^{-1} (if methanol is evaporated at the same rate as H_2O) or ~0.1 μmole m^{-2} d^{-1}, which is on the order of 1% of the measured methanol flux. It would take ~1 day for spray-mediated methanol emission to replace methanol within the lowest 5 m of the atmosphere – a timescale much longer than the eddy covariance averaging period.

We also consider the competing case of spray removing methanol from the atmosphere. The upper limit effect of this can be demonstrated by assuming all spray droplets reach methanol saturation (from an initial concentration of zero) with the atmosphere before falling back to the ocean. At a wind speed of 20 m s^{-1}, the total mass concentration of sea spray is on the order of 1 g of spray/m^3 of air [26]. During HiWinGS, the equilibrium methanol concentration with the atmosphere (HC_a) is on the order of 100 μmole/m^3 of water. Thus 1 g of spray/m^3 of air can take up a maximum of 0.1 nmole of methanol/m^3 of air. This is two orders of magnitude lower than the actual atmospheric methanol concentration. From these calculations, droplet capacity appears to be a limitation to any spray-mediated methanol transfer. The seawater concentration as well as solubility of acetone are lower than those of methanol and we expect the effect of spray on acetone to be even less.

Key uncertainties in the estimations above include the spray source function and the size distribution of the spray droplets. Clearly, further measurements in high winds are needed to more accurately constrain the effect of sea spray on airside gas transfer.

4. Conclusions

In this contribution, we first presented reprocessed methanol and acetone transfer velocities from the HiWinGS cruise. Both transfer rates are close to COARE predictions at wind speeds less than 15 m s^{-1}. At higher wind speeds, measured methanol and acetone transfer velocities appear to be lower than predicted, with the more soluble methanol showing a greater deviation. From the difference between methanol and acetone transfer, we estimated the waterside "zero bubble" transfer velocity, which is fairly close to the COARE predictions for interfacial transfer. We compared the AMT-22 and the HiWinGS cruise in methanol and sensible heat transfer. At wind speeds below 15 m s^{-1}, measurements from the two cruises demonstrate good agreement for both scalars. Above 20 m s^{-1}, measured sensible heat transfer during HiWinGS is lower than the model prediction, qualitatively similar to the behavior of methanol. We crudely estimated the order of magnitude effect of sea spray on methanol transfer, which appears to be small. The reasons for the low gas and sensible heat transfer rates at high wind speeds during HiWinGS remain to be explained.

References

[1] Yang M *et al* 2014 *Atmos. Chem. Phys.* **14** 7499
[2] Liss P S and Slater P G 1974 *Nature* **247** 181
[3] Jurado E *et al* 2004 *Environ. Sci. Technol.* **38** 5505
[4] Lenschow D H *et al* 1981 *J. Geophys. Res.* **86** 7291
[5] Zafiriou O C *et al* 1980 *Geophys. Res. Lett.* **7** 341
[6] Beale R *et al* 2013 *J. Geophys. Res.* **118** doi:10.1002/jgrc.20322

[7] Coburn S *et al* 2014 *Atmos. Meas. Tech.* **7** 3579
[8] Blomquist B *et al* 2006 *Geophys. Res. Lett.* **33** doi:10.1029/2006GL025735
[9] Yang M *et al* 2011 *J. Geophys. Res.* **116** doi:10.1029/2010JC006526
[10] Fairall C W *et al* 2011 *J. Geophys. Res.* **116** doi:10.1029/2010JC006884
[11] Bell M M *et al* 2012 *J. Atmos. Sci.*, **69** 3197
[12] Zhang J A *et al* 2008 *Geophys. Res. Lett.* **35** doi:10.1029/2008GL034374
[13] Andreas E L *et al* 1995 *Bound.-Layer Meteor.* **72** 3
[14] Makin V K 1998 *J. Geophys. Res.* **103** 1137
[15] McInnes L M *et al* 1994 *J. Geophys. Res.* **99** 8257
[16] Sievering H *et al* 1991 *Atmos. Environ.* **25A** 1479
[17] Faloona I *et al* 2010 *J. Atmos. Chem.* **63** 13
[18] Yang M *et al* 2013 *Atmos. Chem. Phys.* **13** 6165
[19] Yang M *et al* 2013 *Proc. Natl. Acad. Sci.* **110** doi:10.1073/pnas.1317840110
[20] Yang M *et al* 2014 *J. Geophys. Res. Oceans* **119** doi:10.1002/2014JC010227
[21] Edson J *et al* 1998 *J. Atmos. Oceanic Technol.* **15** 547
[22] Beale R *et al* 2011 *Analytica Chimica Acta* **706** 128
[23] Edson J B *et al* 2013 *J. Phys. Oceanogr.* **43** 1589
[24] Fairall C W *et al* 1996 *J. Geophys. Res.* **101** 3747
[25] Andreas E L 2010 *J. Phys. Oceanogr.* **40** 608
[26] Fairall C W *et al* 2009 *J. Geophys. Res.* **114** doi:10.1029/2008JC004918

IOP Conf. Series: Earth and Environmental Science **35** (2016) 012012 doi:10.1088/1755-1315/35/1/012012

Air-sea CO_2 fluxes measured by eddy covariance in a coastal station in Baja California, México

L Gutiérrez-Loza[1] **and F J Ocampo-Torres**

Departamento de Oceanografía Física, Centro de Investigación Científica y Educación Superior de Ensenada, Carretera Enseanada–Tijuana No. 3918, Zona Playitas, BC, México

E-mail: `lugutier@cicese.edu.mx`

Abstract. The influence of wave-associated parameters controlling turbulent CO_2 fluxes through the air–sea water interface is evaluated in a coastal region. The study area, located within the Todos Santos Bay, Baja California, México, was found to be a weak sink of CO_2 with a mean flux of -1.32 μmol m^{-2}s^{-1}. The low correlation found between flux and wind speed ($r = 0.09$), suggests that the influence of other forcing mechanisms, besides wind, is important for gas transfer modulation through the sea surface, at least for the conditions found in this study. In addition, the results suggest that for short periods where an intensification of the wave conditions occurs, a CO_2 flux response increases the transport of gas to the ocean.

1. Introduction

Gas transfer at the sea surface is one of the most important factors regarding global climate and long-term climate changes. Despite its importance, there is still great uncertainty in how to parametrize these processes in order to include them in global climate models; this uncertainty exposes the need to increase our knowledge of gas transfer controlling mechanisms. Recently, the importance of carbon fluxes in the coastal zone have been recognized as a major piece of the global carbon cycle [1]; yet estimates of CO_2 fluxes in the coastal zone are not well constrained; a wide range of values has been reported in the literature [2]. It is therefore essential to understand and accurately account for the factors regulating these fluxes in order to accurately estimate their contribution to the ocean and to the global carbon budget [3].

Under calm conditions, gas exchange through the ocean surface occurs by molecular diffusivity due to the difference in gas concentrations between air and water phases. Some studies, however (see e.g., [4],[5],[6],[7]), have shown that turbulent processes in the atmospheric and oceanic boundary layers tend to enhance the efficiency of gas transfer by modifying the behavior of the diffusive layer. This efficiency can be represented by a resistance of the surface and expressed by a transfer velocity. Thus, air-sea CO_2 exchange can be parametrized as,

$$F_{CO_2} = k_{CO_2} K(\Delta pCO_2) \tag{1}$$

where F_{CO_2} is the air-sea flux, k_{CO_2} is the transfer velocity, K is the solubility of CO_2 and ΔpCO_2 is the difference between the partial pressure of CO_2 in the water and in the atmosphere [8].

[1] To whom any correspondence should be addressed.

Content from this work may be used under the terms of the Creative Commons Attribution 3.0 licence. Any further distribution of this work must maintain attribution to the author(s) and the title of the work, journal citation and DOI.
Published under licence by IOP Publishing Ltd

Usually, K and ΔpCO_2 can be determined with precision; it is in the expression of k_{CO_2} where we find the greatest uncertainty when estimating F_{CO_2} [9]. In the ocean, k is commonly expressed as a wind speed function, but it is well known that many other processes including a variety of physical and biogechemical factors can be important when estimating gas fluxes through the sea surface. Further efforts have to be made to understand and include the physical forcings in these parameterizations to achieve closure of the global carbon budget; furthermore, a reduction of the uncertainties in the coastal ocean flux estimates is necessary to be able to extrapolate our understanding of carbon dynamics from local to global scales [3].

The coastal ocean covers only 7% of the total ocean surface [10] but is responsible for up to 30% of the oceanic primary production, 30% to 50% of inorganic carbon and close to 80% of organic carbon burial in sediments [11]. These characteristics, in addition to the physical processes, make the coastal regions very active and heterogeneous ecosystems. The fluxes in these areas are not well represented by the mentioned parametrizations and the lack of high spatial and temporal resolution data hinders the inclusion of the global coastal regions in climate models and leaves open the question of whether these regions are sinks or sources of CO_2.

Previously, Reimer *et al.* [12] compared the CO_2 fluxes at two sites in Todos Santos Bay (TSB), reporting larger fluxes in the intertidal zone than in a site $\sim$ 3 km offshore. Though they carried out different techniques at the two measurement sites, and direct measurements were made only in the intertidal zone, they attribute the differences in the estimated fluxes to the characteristics of the physical forcings at each site.

In this study we present the results of a continuous measurement campaign from which we obtained a full year of high quality data for air-sea CO_2 fluxes in a coastal region. Based on this information we aim to evaluate the influence of wave-associated parameters as controlling mechanisms on turbulent CO_2 fluxes through the air-sea interface. CO_2 and H_2O fluxes were estimated using the eddy covariance method, described in section 2; the main results of this work are presented in section 3, where the relationship between the fluxes and physical parameters such as wind and the wave field is discussed; and in section 5, conclusions and plans for future work are presented.

2. Data and methods

Todos Santos Bay is a small bay ($\sim$ 180 km^2) in the Baja California west coast, approximately 100 km south of the Mexican–U.S. border. The basin is connected to the Pacific Ocean through two segments to the north and south of Todos Santos Island (TSI). The bay has a depth of 50 m in about 80% of its area [13]. TSB lies in a region where upwelling occurs year round, with March–August the months of strongest and more sustained events [14]. The California Current has a strong influence on the thermohaline characteristics of TSB. A sea–land breeze system [15] is one of the main characteristics of TSB, with strong winds from the sea during the day and lower intensities from land during the night.

The study area, known as Punta Morro, is located at the northwest of TSB (figure 1), where direct measurements of CO_2 and water vapor were carried out in a coastal station from May 2014 to April 2015. The measurement tower (figure 2), located on the shoreline at $31°51'41''$N, $116°40'07''$W, was instrumented with two open-path gas analyzers (LI-7500, LI-COR Biosciences) and a sonic anemometer (R3-100 Professional 3D Anemometer, Gill Instruments), all at a height of 13 m above the mean sea level, with a sampling rate of 20 Hz and, given the accessibility of the place, all were powered directly from the electrical network. We only use data retrieved from one of the two gas analyzers because of one instrument failure. In order to reduce the optical contamination, a cleaning system was installed in the functional gas analyzer. Hourly wave data were recorded at 2 Hz using an acoustic Doppler current profiler (Workhorse Sentinel ADCP, Teledyne RD Instruments) deployed at a depth of 10 m and 350 m away from the tower at $31°51'39.6''$N, $116°40'20.28''$W.

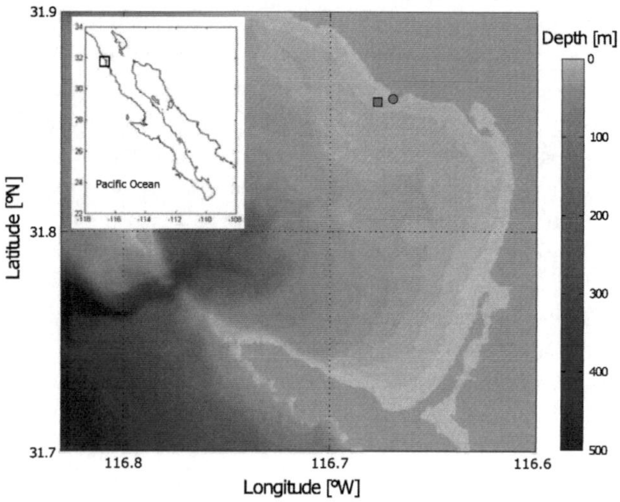

Figure 1. Map of the study site at Todos Santos Bay, Baja California, México. The symbols show the location of the ADCP (■) and the measurement tower (●).

Figure 2. Measurement tower located on the shoreline and instrumented with a sonic anemometer and two open-path gas analyzers.

The CO_2 and water vapor fluxes (hereinafter FCO_2 and FH_2O, respectively) were estimated using the eddy covariance (EC) method [16], which has become the primary method for the estimation of turbulent fluxes in terrestrial and coastal applications. Estimating gas fluxes through EC in the coastal zone allows several timescales to be included in the estimation and resolution of the temporal variability induced by the tidal, diurnal and seasonal cycles [10]. On the other hand, the method is challenging and several corrections are needed to accomplish the physical and theoretical requirements [16]. The EC technique is a non-invasive method that allows gas flux estimations based on measurements of the covariance between fluctuations in the vertical wind velocity and the gas mixing ratio with a high temporal resolution [17]. The general equation describing the flux (F) is:

$$F = \overline{\rho_a w' s'} \tag{2}$$

where ρ_a is the air density [kg m^{-3}], w is the vertical component of the wind speed [m s^{-1}] and s is the mixing ratio of CO_2 and water vapor, with the primes indicating that these values are the fluctuation about their respective means and the overbar indicating the temporal average.

For this study and based on a co-spectral analysis, the averaging period was chosen to be 15 min for the flux estimation and the turbulent fluctuations were obtained through Reynold's decomposition. Quality control procedures were applied to the data before estimating the fluxes. These include de-spiking [18], coordinate rotation [19], sonic corrections [20], atmospheric stability analysis and estimation of mixing ratios as proposed by Sahlée *et al.* [21] to discard the effect of temperature and moisture fluctuations in the FCO_2 estimations. Only on-shore wind directions were taken into account.

The estimated fluxes through the EC method are representative of a certain area, referred to as the footprint; this area depends on parameters such as sensor height, sea surface roughness, atmospheric stability and wind speed. Here we estimated the footprint following Kormann and Meixner [22] and found that in 60% of the cases, under neutral atmospheric conditions, 70%

of the total FCO_2 occurred between the shoreline and 400 m offshore in the predominant wind direction; under unstable conditions (27% of the cases) 45% of the total flux came from the same area close to the shore. On the other hand, the footprint values found under stable conditions were too large (several kilometres) and the FCO_2 under these conditions ($z/L > 1$) were not taken into account.

Applying the SWAN wave model (see [23] and [24]) and considering the directional wave spectrum from the ADCP as a boundary condition, we simulate the wave field over the area covering from the measurement tower to the ADCP (350 m offshore) location. In addition to the integral parameters obtained, the wave steepness and the energy dissipation due to wave breaking were estimated.

3. Results

Wind speed conditions were low to moderate during the entire year with an average value of 4 m s^{-1} and oncoming predominantly from the northwest (figure 3), which is known to induce upwelling [12],[25]. The maximum significant wave height (H_s) recorded was 2.7 m and the average value for the whole year was 0.8 m with the prevailing direction from the southwest due to depth refraction. The description of the environmental parameters is shown in table 1.

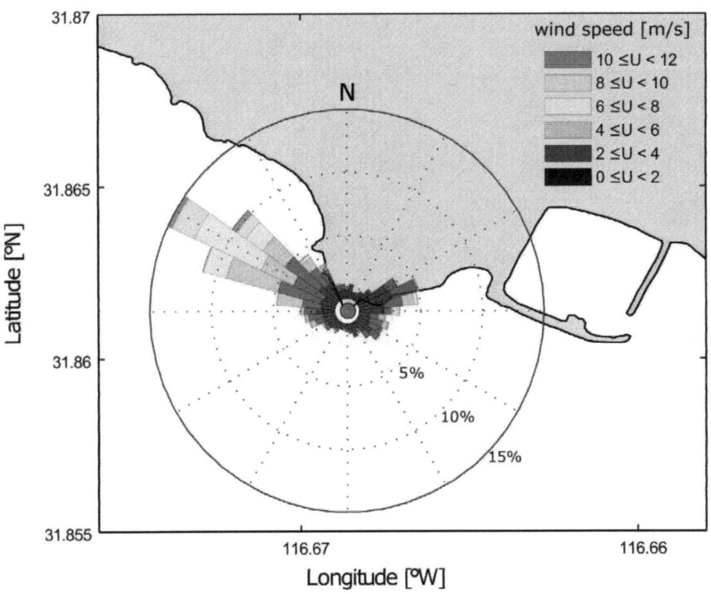

Figure 3. Wind rose at Punta Morro station. The bars indicate the oncoming wind direction, the colors the wind speed and the concentric circles indicate the frequency of occurrence of the data.

Table 1. Environmental parameters measured at Punta Morro from May 2014 to April 2015.

Parameter	Mean ± STD	Min.–Max.	Direction
Wind speed [m s^{-1}]	4.02 ± 2.48	$0.06 - 11.78$	NW
H_s [m]	0.80 ± 0.22	$0.39 - 2.69$	SW
Air Temp. [°C]	20.53 ± 3.59	$8.87 - 32.69$	-
Water Temp. [°C]	16.77 ± 1.78	$12.56 - 21.75$	-
CO_2 conc. [mmol/m^3]	12.58 ± 2.71	$6.55 - 19.43$	-
H_2O conc. [mmol/m^3]	804.41 ± 503.04	$0.66 - 2488.3$	-

The study area was found to be a weak sink of CO_2 during the whole period, with a mean flux and standard deviation of -1.32 ± 8.94 μmol m^{-2}s^{-1}. The estimated CO_2 fluxes are in accordance with Liu *et al.* [26] and Borges *et al.* [2], who suggest that on a global scale, coastal regions are weak sinks of carbon. The mean flux and standard deviation found for H_2O were 0.77 ± 1.28 μmol m^{-2}s^{-1}. There is no particular pattern in the behavior of the fluxes, neither for CO_2 or H_2O (figure 4), but a larger scatter of the data is noticeabe during December–April. This behavior is consistent with the increase of significant wave height (figure 5). Nevertheless, the mean values of FCO_2 remain negative and of similar magnitude in both seasons, indicating that the wave field may be having an effect on the transfer but not modifying the average behavior or describing the direction of the flux.

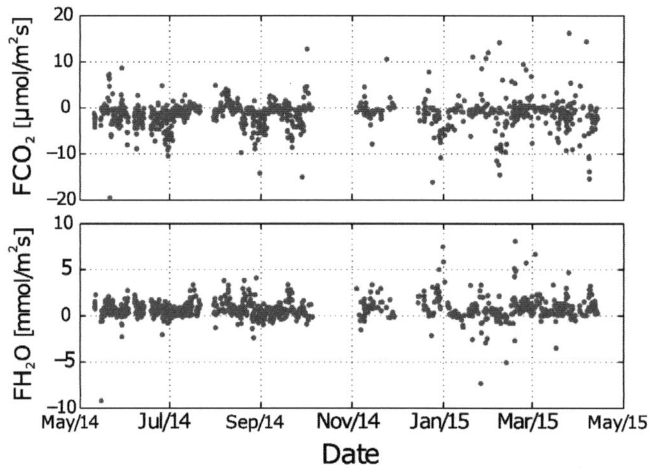

Figure 4. Estimated FCO_2 (upper panel) and FH_2O (lower panel) using 15-min average periods.

Time series of the mean significant wave height, as well as the parameters related to the wave breaking obtained from SWAN model computations are shown in figure 5. An increase of the significant wave height is noticeable for the period October–April with a maximum significant wave height of 2.64 m (figure 5a). This behavior is reflected as an increase in the energy dissipation during that period due to wave breaking (figure 5c), which in coastal regions is known to be the main source of dissipation and for this study was found to be equivalent to the total dissipation. The wave steepness (figure 5b) does not show the same seasonal behavior and maximum values are observed throughout the year, suggesting breaking of smaller waves with less energy dissipation May–October.

To find a relationship between the fluxes and the key physical processes of the region, we represent the fluxes estimated during the entire year as a function of the wind speed and the wave steepness, the latter as an indicative parameter of wave breaking. The results are shown in table 2. The weak correlation between the fluxes and the wind speed suggests that other physical and biological processes may be important factors for gas exchange modulation in coastal waters, but the idea of the wind being a major factor in controlling the fluxes at smaller time scales is not rejected. Similarly, the correlation between fluxes and wave steepness is not enough to establish any relation between both processes and further analysis of the wave field is needed to accomplish a quantitative evaluation of the effect of ocean surface waves on gas exchange in the coastal zone.

IOP Conf. Series: Earth and Environmental Science **35** (2016) 012012 doi:10.1088/1755-1315/35/1/012012

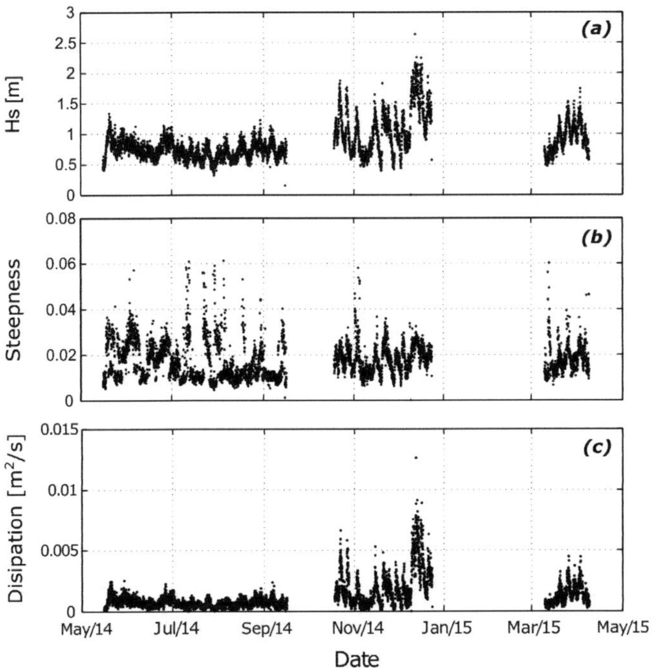

Figure 5. Spatial averages for (*a*) significant wave height (H_s), (*b*) wave steepness obtained as the ratio between H_s and the wavelenght (L), and (*c*) dissipation due to wave breaking.

Table 2. Statistical description of the relationship between fluxes and the physical parameters.

	Correlation (r)	Lower limit*	Upper limit*	*p*-value**
FCO_2 vs. wind speed	−0.09	−0.15	−0.03	0.0061
FH_2O vs. wind speed	0.32	0.26	0.38	< 0.0001
FCO_2 vs. steepness	−0.16	−0.24	−0.08	0.0002
FH_2O vs. steepness	0.16	0.08	0.24	0.0001

* 95% confidence interval.
** All correlations were statistically significant at $p< 0.05$.

Wave steepness is used as a indicative parameter of wave breaking but in this case we found that it does not explain gas flux behavior. This is because fluxes are not affected directly by the presence of breaking waves but rather by the magnitude of the turbulent processes modifying the air-sea interface at both water and air boundary layers. This explains the low correlation between the fluxes and the wave steepness presented in table 2.

Two 7-day periods with high dissipation rates were selected for further analysis. Good agreement between FCO_2 and the significant wave height was found for these periods with a correlation coefficient of $r = -0.56$ and statistical significance (figure 6).

4. Conclusions

A full year of high-quality coastal CO_2 air–sea flux data are presented. The observation area of TSB was found to be a weak sink of CO_2 for the entire year, based on measurements made between May 2014 and April 2015, with a mean CO_2 flux of -1.32 μmol $m^{-2}s^{-1}$. The low correlation between CO_2 flux and wind speed ($r = -0.09$) suggests that the influence of other forcing mechanisms is important for the transfer modulation through the sea surface, at least for the conditions observed during this study.

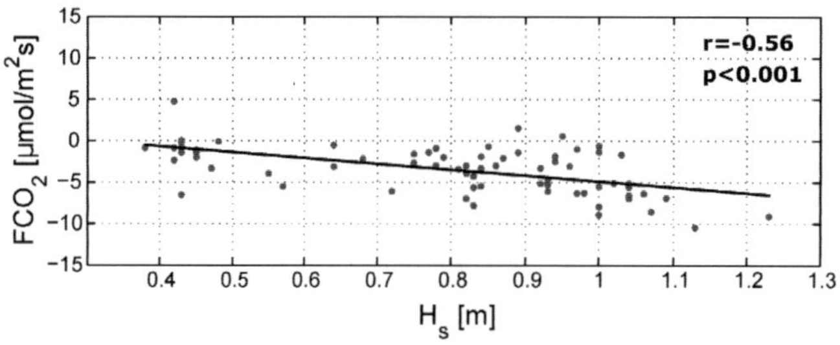

Figure 6. FCO_2 vs. significant wave height (H_s) during two 7-day periods with high dissipation due to wave breaking.

An increase in the wave field was observed for the period December–April, apparently causing scatter in both CO_2 and H_2O fluxes even when no seasonality was found in the flux behavior; this indicates that the wave field may have an effect on gas transfer but did not modify the average behavior.

The dissipation rate caused by the breaking waves was estimated using the SWAN waves model as a measurement of the turbulence caused by the wave field. Good agreement ($r = -0.56$) between the CO_2 flux and significant wave height (H_s) is evident for two periods during which the average dissipation rate increased. The results suggest that for short periods where an intensification of the wave conditions occur, a CO_2 flux response increases the flux into the ocean. This response was not observed for the entire period, which indicates that the effect of other biochemical and physical processes are also important. Further analysis will be needed to better understand the conditions under which the wave field plays a major role in gas flux behavior and to what extent.

Acknowledgments

We thank the CICESE Physical Oceanography Department and the 7th International Symposium on Gas Transfer at Water Surfaces organizing committee for their financial support. We thank J Martínez, R Alcaraz and E Rivera for technical help related to experimental set-up and maintenance of instruments. Finally, we thank P Osuna for his advice and support, and the rest of the CICESE wave research group for their contributions to this work. This study is part of the "Sea surface roughness as air-sea interaction control" (RugDiSmarCTRLo-a)CONACyT project (155793).

References

[1] IPCC *Climate Change 2013: The physical Science Basis: Contribution of Working Group I to the Fifth Assessment Report of the Intergovernmental Panel on Climate Change* Ed. T F Stocker *et al.* (New York: Cambridge University Press)

[2] Borges A V, Delille B and Frankignoulle M 2005 Budgeting sink and sources of CO_2 in the coastal ocean: diversity of ecosystems counts *Geophys. Res. Lett.* **32**(14), doi:10.1029/2005GL023053

[3] Bauer J E, Cai W–J, Raymond P A, Bianchi T S, Hopkinson C S and Regnier P A G 2013 The changing carbon cycle of the coastal ocean *Nature* **504**(7478) 61–70, doi:10.1038/nature12857

[4] Ocampo-Torres F J, Donelan M A, Merzi N and Jia F 1994 Laboratory measurements of mass transfer of carbon dioxide and water vapour for smooth and rough flow conditions *Tellus* B **46**(1) 16–32

[5] McGillis W R, *et al.* 2004 Air–sea CO_2 exchange in the equatorial Pacific *J. Geophys. Res.* **109**(C8), doi:10.1029/2003JC002256

[6] Zappa C J, McGillis W R, Raymond P A, Edson J B, Hintsa E J, Zemmelink H J, Dacey J W H and Ho D T 2007 Environmental turbulent mixing controls on air–water gas exchange in marine and aquatic systems *Geophys. Res. Lett.* **34**(10), doi:10.1029/2006GL028790

[7] Jähne B, Münnich K O, Bösinger R , Dutzi A, Huber W and Libner P 1987 On the parameters influencing air-water gas exchange *J. Geophys. Res.* **92**(C2) 1937–49
[8] Wanninkhof R, Asher W E, Ho D T, Sweeney C and McGillis W R 2009 Advances in quantifying air-sea gas exchange and environmental forcing *Annu. Rev. Mar. Sci.* **1** 213–44, doi:10.1146/annurev.marine.010908.163742
[9] Mørk E T, Sørensen L L, Jensen B and Sejr M K 2014 Air-sea CO_2 gas transfer velocity in a shallow estuary *Bound.-Lay. Meteorol.* **151** 119–38, doi: 10.1007/s10546-013-9869-z
[10] Polsenaere P *et al.* 2012 Spatial and temporal CO_2 exchanges measured by Eddy Covariance over a temperate intertidal flat and their relationship to net ecosystem production *Biogeosciences* **9** 249–68, doi:10.5194/bg-9-249-2012
[11] Jahnke R A 2010 Global synthesis *Carbon and nutrient fluxes in continental margins* Eds. K K Liu, L Atkinson, R Quiones and L Talaue–McManus (Springer) p 597.
[12] Reimer J J, Vargas R, Smith S V, Lara–Lara R, Gaxiola–Castro G, Hernández–Ayón J M, Castro A, Escoto-Rodriguez M and Martínez–Osuna J 2013 Air–sea CO_2 fluxes in the near–shore and intertidal zones influenced by the California Current *J. Geophys. Res.* **118**(10) 4795–810, doi:10.1002/jgrc.20319
[13] Mateos E, Marinone S G, Parés–Sierra A 2009 Towards the numerical simulation of the summer circulation in Todos Santos Bay, Ensenada, B.C, Mexico *Ocean Model.* **27** 107–12
[14] Feely R A, Sabine C L, Hernandez–Ayon J M, Ianson D and Hales B 2008 Evidence for upwelling of corrosive "acidified" water on to the continental shelve *Science* **320**(5882) 1490–92, doi: 10.1126/science1155676
[15] Reyes S y Pars A 1983 Analisis de las componentes principales de los vientos superficiales sobre la bahia de Todos Santos *Geof. Int.* **22**(2) 179-203
[16] Baldocchi D D, Hincks B B and Meyers T P 1988 Measuring biosphere–atmosphere exchanges of biologically related gases with micrometeorological methods *Ecology* **69**(5) 1331–40
[17] Aubinet M *et al.* 2000 Estimates of the annual net carbon and water exchange of forests: the EUROFLUX methodology *Adv. Ecol. Res.* **30** 113-75
[18] Vickers D and Mahrt L 1997 Quality control and flux sampling problems for tower and aircraft data *J. Atmos. Oceanic Tech.* **14**(3) 512–26
[19] Kaimal J C and Finnigan J J 1994 Sensors and techniques for observing the boundary layer *Atmospheric boundary layer flows: their structure and measurement* (New York: Oxford University Press)
[20] Schotanus P, Nieuwstadt F T M and De Bruin H A R 1983 Temperature measurement with a sonic anemometer and its application to heat and moisture fluxes *Bound.-Lay. Meterol.* **26**(1) 81–93
[21] Sahlée E, Smedman A S, Rutgersson A and Högström U 2007 Spectra of CO_2 and water vapour in the marine atmospheric surface layer *Bound.-Lay. Meteorol.* **126**(2) 279–95
[22] Kormann R and Meixner F X 2001 An analytical footprint model for non-neutral stratification. *Bound.-Lay. Meteorol.***99**(2) 207–
[23] Booij N, Ris R C and Holthuijsen L H 1999 A third–generation wave model for coastal regions 1.Model description and validation *J. Geophys. Res.* **104**(C4) 7649–66
[24] Ris R C, Holthuijsen L H and Booij N 1999 A third–generation wave model for coastal regions 2.Verification *J. Geophys. Res.* **104**(C4) 7667–81
[25] Pérez–Brunius P, López M, Parés–Sierra A and Pineda J 2007 Comparison of upwelling indices off Baja California derived from three different wind data sources. *CalCOFI Reports* **48** 204–214
[26] Liu K K, Atkinson L, Chen C T A, Gao S, Hall J, Macdonald R W, Talaue McManus L and Quiñoes R 2000 Exploring continental margin carbon fluxes on global scale *Eos. Trans. AGU* **81**(52) 641–44

IOP Conf. Series: Earth and Environmental Science **35** (2016) 012013 doi:10.1088/1755-1315/35/1/012013

Water surface slope spectra in nearshore and river mouth environments

N J M Laxague, D G Ortiz-Suslow, B K Haus, N J Williams and H C Graber

University of Miami, Rosenstiel School of Marine and Atmospheric Science, Department of Ocean Sciences, 4600 Rickenbacker Causeway, Miami, FL 33149, USA

E-mail: nlaxague@rsmas.miami.edu

Abstract. With the ever-growing interest in satellite remote sensing, direct observations of short wave characteristics are needed along coastal margins. These zones are characterized by a diversity of physical processes that can affect sea surface topography. Here we present connections made between ocean wave spectral shape and wind forcing in coastal waters using polarimetric slope sensing and eddy covariance methods; this is based on data collected in the vicinity of the mouth of the Columbia River (MCR) on the Oregon–Washington border. These results provide insights into the behavior of short waves in coastal environments under variable wind forcing; this characterization of wave spectra is an important step towards improving the use of radar remote sensing to sample these dynamic coastal waters. High wavenumber spectral peaks are found to appear for $U_{10} > 6$ m/s but vanish for $\tau > 0.1$ N/m^2, indicating a stark difference between how wind speed and wind stress are related to the short-scale structure of the ocean surface. Near-capillary regime spectral shape is found to be less steep than in past observations and to show no discernable sensitivity to wind forcing.

1. Introduction

In the study of physical processes and management of resources in coastal and nearshore environments, satellite radar remote sensing techniques are important and widely used tools. These techniques rely on back-scattering of transmitted radar signals by short ocean waves. Scattering models [1] depend upon the short ocean wave spectrum (i.e., decimeters and smaller — the range for which capillary effects become significant) in the determination of the normalized radar cross section (NRCS), or σ_0 — a quantity of primary importance for the retrieval of meteorological and oceanographic parameters from satellite images. The ability to accurately collect these parameters remotely is a powerful one — sea surface waves are of crucial significance to all manner of interactions between the atmosphere and ocean. Heat and gas fluxes are dramatically enhanced by the presence of ocean surface waves [2], holding implications for areas of study ranging from extreme storms to climatic processes. The transfer of momentum from the atmosphere into the ocean via wind forcing is inextricably connected to wind-wave growth, upper ocean mixing, and mass transport.

The vertical transfer of horizontal momentum from the atmosphere to the ocean (i.e., wind stress) is generally quantified using the eddy covariance method [3]. This relates the turbulent fluctuations in the wind velocity field to a characteristic wind velocity at the surface. Over water, it has long been established that these turbulent fluctuations are related to the small scale roughness features on the surface. Any process that affects the small scale surface waves ($\mathcal{O}(0.01-0.1)$ m) will modulate the atmospheric response to the surface [4].

Content from this work may be used under the terms of the Creative Commons Attribution 3.0 licence. Any further distribution of this work must maintain attribution to the author(s) and the title of the work, journal citation and DOI.
Published under licence by IOP Publishing Ltd

There are a number of pitfalls associated with observing these short waves (even the 0.1-m waves are typically not sampled by buoys), especially outside of a controlled laboratory environment. Traditional time series observations are unable to properly isolate the effects of advection and Doppler-shifting of short waves by long waves [5]. For this reason, techniques that are able to retrieve synoptic, spatial measurements of wave structure are preferred for the study of short ocean waves [6]. The polarimetric slope sensing (PSS) technique is an optical method designed to accomplish such a task [7]; it has been used successfully to acquire short-scale wave structure from aboard a moving vessel [8, 9]. This has the advantage of adapting this relatively new field-ready technology to heterogeneous ocean surfaces where both spatial and temporal variability are first order concerns.

Figure 1. Images taken from and of the R/V *Point Sur*, clockwise from top left: view of the Washington coast from MCR; bow-mounted meteorological mast; starboard bow-mounted polarimetric and infrared cameras; starboard bow-mounted ultrasonic distance meter (UDM).

For this study, short wave and flux measurements aboard a research vessel were used. Field observations of wind stress and short scale wave slope were made in the vicinity of the Columbia River mouth aboard the R/V *Point Sur*. These observations were performed as part of the Office of Naval Research-sponsored Riverine and Estuarine Transport Experiment (RIVET) II. An over-arching goal of this multi-faceted campaign was to provide *in situ* observations of coastal wave-current-wind interaction that would improve efforts to use remote sensing platforms for sampling these dynamic regions. The work presented here furthers this aim by providing short wave observations fundamental to the acquisition methods of remote sensing sampling techniques.

2. Methods

2.1. Field observations: Mouth of the Columbia River

Observations were made at the mouth of the Columbia River (MCR) along the Oregon–Washington border from 25 May to 15 June 2013. The peak wave directions observed by NDBC buoy #46243 during the experimental time frame show waves generally incident from the west-northwest, with typical peak periods ranging from 7 to 12 s. Surface slope and wind stress measurements were made from the bow of the 135 foot R/V *Point Sur* (figure 1). A sonic anemometer was fastened atop a meteorological mast mounted (2.3 m from deck) on the bow providing three-dimensional wind velocities 8.1 m above the mean water level (MWL). A polarimetric camera was mounted off the starboard bow (figure 1) and oriented such that it was imaging an ≈ 1.56 m^2 region forward of the ship's wake zone. The vessel's angular and linear accelerations were logged at 20 Hz via a Systron-Donner six-axis motion package, which was located adjacent to the sensor suite on the ship's bow. This was used to motion-correct the flux measurements and to properly rectify the polarimeter frames. The following sections describe the methods used in retrieving the shipboard measurements of the wind stress and short wave slope spectra.

2.2. Wind stress: Eddy covariance

Given here is a condensed version of the estimation of wind stress via eddy covariance that is directly relevant to this study. The total stress vector on the ocean surface is generally taken as the covariance of the turbulent along- and across-wind velocities (u' and v') with the vertical wind component (w'),

$$\vec{\tau}_{wind} = -\rho_{air}\left[\langle u'w'\rangle\hat{x} + \langle v'w'\rangle\hat{y}\right] \tag{1}$$

where ρ_{air} is the air density and a prime indicates a turbulent quantity in a Reynolds decomposition (i.e., $\langle u'\rangle = \langle v'\rangle = \langle w'\rangle = 0$). The brackets indicate that a time average of the contained quantity has been computed over some appropriate interval. Simultaneous measurements of vessel motion enabled the 10-Hz winds measured from the sonic anemometer to be corrected for the vessel accelerations and translation [10]. Rejection criteria included averaging intervals with large vessel heading deviations ($\geq 40°$) and winds coming from outside of $\pm 40°$off the bow. Critical to the eddy covariance technique is the use of an appropriate averaging window, which can vary from study-to-study and platform-to-platform. Generally, intervals between 15 and 30 minutes are used in deep water conditions, where spatial homogeneity can be assumed over the sampling interval. A five-minute window was chosen for this study [11] due to the combination of the spatial heterogeneity of the wave and wind stress field observed at MCR and the area covered by the vessel ($\sim$ 1200 m per interval). It is expected that the short surface waves respond most directly to the smaller-scale eddies in the atmospheric boundary layer, which are fully resolved over this time frame. The U_{10} reported in this work is the 10-m, neutral wind speed calculated using the eddy covariance techniques, while the wind stress, τ, comes directly from the along- and across-wind covariance components of the Reynolds stress (equation 1).

2.3. Short waves: Polarimetric slope sensing

Following the work of Zappa *et al.* [2008], one may obtain the two-dimensional slope field $\vec{\sigma}(x,y)$ from a triplet of simultaneously-acquired images representing light intensities at 0°, 45°, and 90°polarization. The imaged area of the water surface spans 1.25 m by 1.25 m, with a minimally-resolvable (Nyquist) wavenumber of 1600 rad/m (corresponding to 3.9 mm). The primary simplifying assumptions for this work are that the incident light be unpolarized and the subsurface, upwelling light may be neglected. Multiple sky-looking acquisition runs showed that the light incident upon the surface was, in general, in an unpolarized state for the cases considered here. For the work presented here (as well as in Zappa *et al.* [2008]), upwelling light was taken to be negligible in general. The polarimeter frames were projectively transformed onto the ocean surface in order to maintain a uniform scale across the image under a time-varying camera look angle [9].

$$\begin{aligned} S_0 &= I_0 + I_{90} \\ S_1 &= I_0 - I_{90} \\ S_2 &= 2 \cdot I_{45} - S_0 \end{aligned} \tag{2}$$

I_0, I_{45}, and I_{90} are the light intensities at 0°, 45°, and 90°, respectively. The Stokes parameters S_1 and S_2 are directly converted into the polarization orientation, from which follows the angle of intersection between the reflection and image planes. The tangent of this angle yields the slope in the cross-look direction.

$$\varphi = \frac{1}{2}\left[\pi + \tan^{-1}\left(\frac{S_2}{S_1}\right)\right], \qquad \sigma_y = \tan(\varphi) \tag{3}$$

In order to compute the slope in the along-look direction, the degree of linear polarization (DOLP) is generated using all three Stokes parameters (equation 4).

$$DOLP = \sqrt{\frac{S_1^2 + S_2^2}{S_0^2}} = \frac{r_\parallel^2 - r_\perp^2}{r_\parallel^2 + r_\perp^2} \tag{4}$$

$$r_\parallel = \frac{\tan(\theta_i - \theta_t)}{\tan(\theta_i + \theta_t)}, \quad r_\perp = \frac{\sin(\theta_i - \theta_t)}{\sin(\theta_i + \theta_t)} \tag{5}$$

$$\theta_i = n \sin(\theta_t), \quad \sigma_x = \tan(\theta_i) \tag{6}$$

The DOLP may alternatively be represented in terms of the Fresnel reflection coefficients (equation 4; defined in equation 5), which themselves depend upon the angle of incidence and the index of refraction n of water. Equation 6 describes the final step in producing a slope component from the observed angle of incidence. For this analysis, $n = 1.33$ was chosen based on an average water temperature and salinity of 14°C and 10 ppt, respectively; analysis of the *Point Sur* flow-through temperature and salinity data demonstrated that n should not have deviated more than 1% from this value for the duration of the experiment [12]. The focus of this work will be on the spectral properties of the time-averaged two-dimensional slope fields. Each scalar slope field $\|\vec{\sigma}(x,y)\|$ is subjected to a two-dimensional Fourier transform, yielding the two-dimensional slope spectrum $P(k_x, k_y)$. This may be integrated over direction to produce the omnidirectional slope spectrum $P(k)$. By multiplying $P(k)$ by k, one extracts the non-dimensional, omnidirectional saturation spectrum $B(k)$. The high-wavenumber regimes of this spectrum are emphasized; as such, it is often used to describe the short-wave behavior of the ocean surface. Empirical and theoretical wave spectra make fundamental use of the short-wave peak and post-peak behavior of $B(k)$ in defining spectral shape [13, 14]. $P(k)$ itself may be integrated with respect to wavenumber k to compute mean square slope ($\langle S^2 \rangle$ — literally, the mean of the squares of the slopes), an important statistical quantity used to represent surface roughness.

In the section of results that follows (section 3), references will be made to "inside-inlet" and "outside-inlet" data points. While it is difficult to set a hard line of demarcation between areas considered "inside" and "outside", the offshore edge of the south jetty at 124°6' W seemed to be the most reasonable separation point. From figure 2, one can indeed see that this is very nearly the westernmost magenta acquisition location. The separation of data between inside and outside-inlet groups provides a means of comparing two different physical regimes where major differences in short wave statistics are expected. Sections 3 and 4 also refer to wavenumber regimes. These follow from and correspond to the relative values of the gravity and surface tension terms in the linear dispersion relation. They are: short gravity (37.1 rad/m $< k <$ 112.7 rad/m), gravity-capillary (112.7 rad/m $< k <$ 1173 rad/m), and pure capillary (1173 rad/m $< k <$ 1600 rad/m).

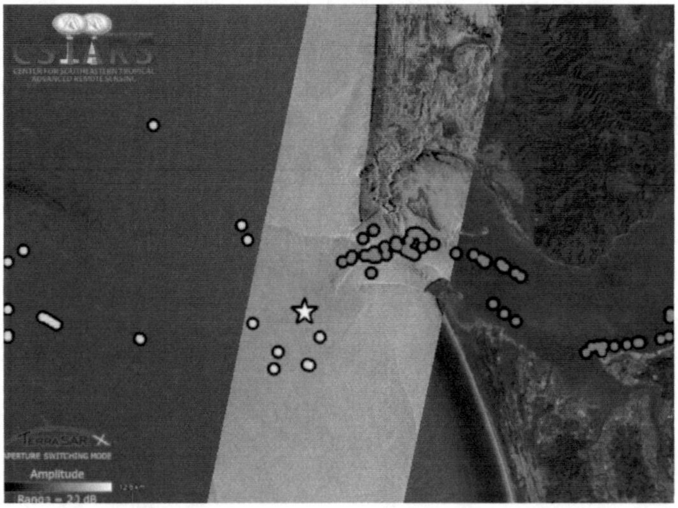

Figure 2. Map of MCR with data acquisition locations marked and a sample TerraSAR-X intensity image overlaid. SAR image was collected on 2 June 2013. Magenta dots represent those runs taken as inside the mouth; cyan dots represent those runs taken as outside the mouth. This color code will remain consistent throughout the figures presented here. The yellow star represents NDBC buoy #46243.

3. Results

For all values of wind speed and wind stress magnitude, k_p outside the inlet is smaller than k_p inside the inlet, with little observed wind sensitivity. The re-processed open-ocean results of [9] (figure 4) are also given for consideration; however, the inside-inlet values appear to be closest to those results. For higher wind speed values, a gravity-capillary regime peak (≈ 500 rad/m) appears for both outside and inside-inlet cases. These values are very close to those obtained from the stereoscopic observational spectra of Yurovskaya *et al.* [6]. The separation between k_p plotted against wind speed and k_p plotted against wind stress manifests itself in the existence of this high-wavenumber peak. This can be seen explicitly in the four spectra presented in figure 3: the example cases all have similar wind speed magnitudes, but those with low wind stress (3a, 3b) show a high-wavenumber peak. For the full body of the data set, figures 4a and 4b appear to mirror one another. As in figure 3, greater scatter in k_p and higher-wavenumber spectral peaks are associated with high wind speed and low wind stress. This disparity will be discussed in greater detail in section 4.

The spectral shape of the near-capillary regime (the high-wavenumber portion of the gravity-capillary regime) is shown in figure 5. The parameter n (i.e., $B(k) \sim k^n$) is plotted against wind speed U_{10} rather than wind stress magnitude τ so that the results from past observations [6, 9] can be shown in their native form in order to avoid invoking a wind stress parameterization. Whereas the spectra from [9] show good agreement with those of [6], those observed for this work are generally steeper, falling off like k^{-2}. None of the spectra shown have a discernable steepening or shallowing trend with wind speed, though there is a regime between $U_{10} = 7$ m/s and $U_{10} = 9$ m/s for which the scatter in outside-inlet spectral fall-off is greater, with some cases exhibiting $k^{-2.5}$ behavior.

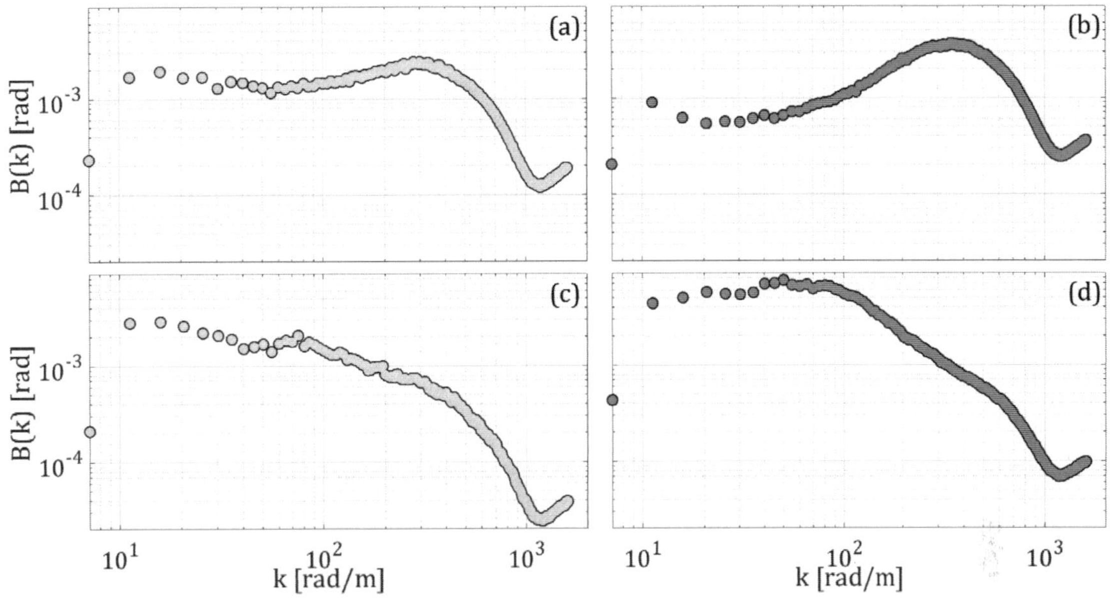

Figure 3. Four selected omnidirectional saturation spectra B(k). Magenta cases are inside-inlet. Cyan cases are outside-inlet. Wind conditions: (a) U_{10} = 7.879 m/s, τ = 0.0117 N/m²; (b) U_{10} = 7.952 m/s, τ = 0.0688 N/m²; (c) U_{10} = 8.548 m/s, τ = 0.139 N/m²; (d) U_{10} = 8.476 m/s, τ = 0.102 N/m².

4. Discussion

Analysis of this Columbia River short wave and wind stress data set provides valuable observations of the shape and scale of the gravity-capillary and near-capillary regimes of the ocean wavenumber spectrum. The initial high value and tapering off of scatter in k_p as wind stress magnitude increases (shown in figure 4b) is an important part of this picture. In times of low wind forcing, feature interactions that might induce wave breaking (wave-wave, wave-geography, etc.) are likely to be significant drivers of variations in the position of the short wave spectral peak and the near-capillary regime shape. As wind forcing magnitude increases, the aforementioned interactions become secondary and the position of k_p moves almost exclusively into the short gravity wave regime. It is important to note that within the river inlet, waves grow over shorter fetches, ranging between 3 and 5 km in the regions sampled. The data shown in figure 4b is a graphical indication that sea states with high-wavenumber spectral peaks may exist in absence of significant wind stress magnitude and in sea states that have evolved over short distances. The spectra shown in figure 3 illustrate the connection between wind stress magnitude and k_p. The fact that "high" (≈ 500 rad/m) wavenumber peaks all but disappear for the upper half of the wind stress magnitude range could be indicative of an increase in the lower-wavenumber (i.e., centimeter and decimeter-scale) features on the ocean surface that are relevant for the atmospheric turbulence used to estimate τ. A cursory look at 4a (i.e., k_p as a function of U_{10}) would not lead to this result, underlining the importance of computing wind stress during investigations of k_p.

In the examination of the near-capillary regime spectral shape (as in figure 5), one sees that the spectra observed here are steeper than those of prior observations [6, 9], though all share a similar, general insensitivity to wind speed. There exists no consensus in the body of capillary wave spectral data that suggests a clear expected behavior. Observational spectra [6, 9] that extend into the capillary or near-capillary regime are sparse, while empirical/parameterized spectra [13, 15] vary greatly in that domain, owing largely to the diverse range of observations and theoretical frameworks invoked in their construction [1, 13, 15].

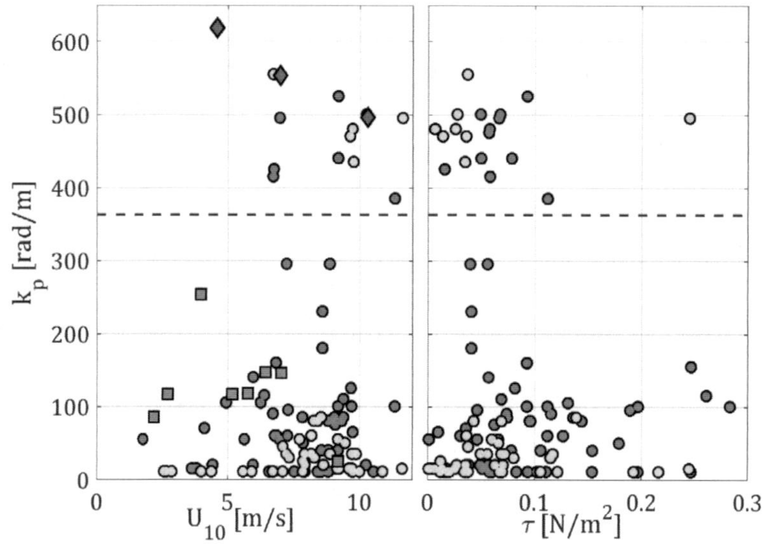

Figure 4. Peak wavenumbers of the omnidirectional saturation spectra, plotted against wind speed magnitude U_{10} (a) and wind stress magnitude τ (b). Magenta dots represent those runs taken as inside the mouth; cyan dots represent those runs taken as outside the mouth. The blue dashed line marks the wavenumber (370 rad/m) at which the gravity-capillary minimum phase speed (0.23 m/s) occurs. Green diamonds represent the peaks of the stereoscopic observational spectra of Yurovskaya *et al.* [6] and orange squares represent the peaks of the polarimetric observational spectra of Laxague *et al.* [9].

5. Conclusion

In summary, these results offer some insight to the subject of coastal air-sea interaction, and not only in the areas previously mentioned (remote sensing, etc.). The disparate behavior of k_p with respect to U_{10} and τ underscores the importance of short ocean waves in turbulent momentum flux. Geographic and spatial considerations likely figured heavily into these observations — the Columbia River mouth is unique in its interaction between highly energetic ocean-like conditions and river-like conditions in such close proximity and at such a grand scale. Background current and long wave effects have not been explicitly considered here, though their impact should be considered in future analysis of the data set. The estimates of near-capillary regime spectral shape do not form a significant result in their own right; however, they represent an important addition to the body of observational ocean surface short wave data. Future study in this vein would include the extension of surface roughness towards existing Columbia River satellite radar remote sensing acquisition, particularly with regard to hydrodynamic wave modulation. The use of these spectral shape measurements along with existing scattering models would allow the authors to independently evaluate NRCS from polarimetric slope data. Finally, it is only natural to dedicate more time and energy to better understanding the complex dynamics impacting short wave directionality. A more rigorous study of the factors affecting it, including wind speed/stress direction, long wave direction and current direction, follows directly from this work.

Acknowledgments
This research was supported by grant #N000141410643 from the Office of Naval Research and grant #SA1207GOMRI005 from BP/the Gulf of Mexico Research Initiative. Thanks are given to all those who worked on and prepared for our portion of the RIVET-II experiment, in particular Mike Rebozo. The authors are grateful to the captain and crew of the R/V *Point Sur* for their hard work and careful execution through conditions ranging from quiet to rough.

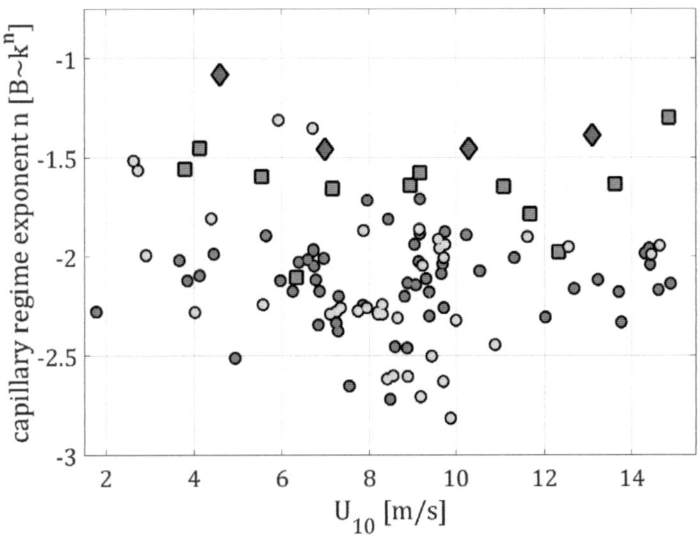

Figure 5. Near-capillary regime spectral shape exponent n (for $B(k) \sim k^n$), plotted against wind speed magnitude U_{10} for comparison with other observational spectra. Magenta dots represent those runs taken as inside the mouth; cyan dots represent those runs taken as outside the mouth. Green diamonds represent the shapes of the stereoscopic observational spectra of Yurovskaya *et al.* [6] and orange squares represent the shapes of the polarimetric observational spectra of Laxague *et al.* [9].

References

[1] Kudryavtsev V, Hauser D, Caudal G and Chapron B *Journal of Geophysical Research: Oceans* **108** ISSN 2156-2202
[2] Frew N M, Bock E J, Schimpf U, Hara T, Hauecker H, Edson J B, McGillis W R, Nelson R K, McKenna S P, Uz B M and Jähne B *Journal of Geophysical Research: Oceans* **109** ISSN 2156-2202
[3] Smith S D 1980 *J. Phys. Oceanogr.* **10** 18
[4] Charnock H 1955 *Q. J. R. Meteorol. Soc.* **81** 1
[5] Hara T, Bock E J and Lyzenga D *Journal of Geophysical Research: Oceans* **99** ISSN 2156-2202
[6] Yurovskaya M, Dulov V, Chapron B and Kudryavtsev V 2013 *Journal of Geophysical Research: Oceans* **118** 4380–4394
[7] Zappa C J, Banner M L, Schultz H, Corrada-Emmanuel A, Wolff L B and Yalcin J 2008 *Measurement Science and Technology* **19** 055503
[8] Zappa C J, Banner M L, Schultz H, Gemmrich J R, Morison R P, LeBel D A and Dickey T 2012 *Journal of Geophysical Research: Oceans (1978–2012)* **117**
[9] Laxague N J, Haus B K, Bogucki D and Özgökmen T 2015 *Journal of Geophysical Research: Oceans* **120** 3140–3156
[10] Anctil F, Donelan M A, Drennan W M and Graber H C 1994 *J. Atmos. Ocean. Technol.* **11** 7
[11] Ortiz-Suslow D, Haus B, Williams N, Laxague N, Reniers A and Graber H 2015 *Journal of Geophysical Research: Oceans* **120** 660–676
[12] Millard R and Seaver G 1990 *Deep Sea Research Part A. Oceanographic Research Papers* **37** 1909–1926
[13] Elfouhaily T, Chapron B, Katsaros K and Vandemark D 1997 *Journal of Geophysical Research-Oceans* **102** 15781–15796
[14] Kudryavtsev V, Makin V and Chapron B 1999 *J. Geophys. Res* **104** 7369
[15] Hwang P A, García-Nava H and Ocampo-Torres F J 2011 *J. Phys. Oceanogr.* **41** 1227–1238

IOP Conf. Series: Earth and Environmental Science **35** (2016) 012014 doi:10.1088/1755-1315/35/1/012014

Enhanced nighttime gas emissions from a lake

E Podgrajsek, E Sahlée and A Rutgersson

Department of Earth Sciences, Air, Water, and Landscape Sciences, Uppsala University, Uppsala, Sweden

E-mail: erik.sahlee@met.uu.se

Abstract. Methane (CH_4) and carbon dioxide (CO_2) are two important greenhouse gases. Previous studies have shown that lakes can be important natural sources of atmospheric CH_4 and CO_2. It is therefore important to monitor the fluxes of these gases between lakes and the atmosphere in order to understand the processes that govern the exchange. Most previous lake flux studies are based on chamber measurements, by using the eddy covariance method, the resolution in time and in space of the fluxes is increased, which gives more information on the governing processes. Eddy covariance measurements at a Swedish lake show that both methane fluxes (FCH_4) and carbon dioxide fluxes (FCO_2) experience high nighttime fluxes for a large part of the data set (largest median $FCH_{4night} \approx 13$ nmol m^{-2} s^{-1} and smallest median $FCH_{4day} \approx 4.0$ nmol m^{-2} s^{-1}, largest median $FCO_{2night} \approx 0.2$ μmol m^{-2} s^{-1} and smallest median $FCO_{2day} \approx 0.02$ μmol m^{-2} s^{-1}, with larger variability during night). For the diel cycle of the CH_4 fluxes it is suggested that water side convection could enhance the transfer velocity, transport CH_4 rich water to the surface, as well as trigger ebullition. The high nighttime CO_2 fluxes could to a large extent be explained with enhanced transfer velocities due to water side convection. If gas fluxes are not measured during nighttime, when water side convection normally is generated, periods of potential high gas flux might be missed and estimations of the total amount of gas released from lakes to the atmosphere will be biased.

1. Introduction

Gas fluxes of carbon dioxide (CO_2) and methane (CH_4) are important parts in the global carbon cycle. The global carbon cycle describes how carbon is stored, processed, and transported between different domains such as the ocean, fossil fuel reservoirs, land, and atmosphere. Freshwater systems, including lakes and rivers, have long been seen as only a funnel transporting carbon from the land domain to the ocean, without processing the carbon. Today it is well know that this is not the case [1, 2]. According to Bastviken *et al.* [2] lakes can outweigh the land sink by 25% taking into account CH_4 release only.

To obtain knowledge on processes affecting the gas fluxes of CO_2 and CH_4 over the lake surface, scientists have started to use the eddy covariance (EC) technique for lake gas flux studies. The EC technique, which is frequently used for land and ocean studies, measures the fluxes continuously without much labor. Instruments measuring the turbulent fluctuations of three different wind components and the turbulent fluctuations of the gas concentration are needed when using the EC technique.

Here we present EC measurements of both CO_2 and CH_4 fluxes from one Swedish lake.

Content from this work may be used under the terms of the Creative Commons Attribution 3.0 licence. Any further distribution of this work must maintain attribution to the author(s) and the title of the work, journal citation and DOI.
Published under licence by IOP Publishing Ltd

2. Methods and site

2.1. Site

The measurement site, Lake Tämnaren, is located in central Sweden, 60°09'N, 17°20'E (figure 1). Tämnaren is a shallow lake with a mean depth of only 1.3 m (maximum depth 2 m) and an area of 38 km^2. The EC tower at Lake Tämnaren was 6 m high with three levels, 1.4, 2.7, and 6 m above ground, equipped with propeller anemometers for wind speed and wind direction (Young, MI, USA) and radiation shielded and ventilated thermocouples for measurements of air temperature. The EC instrumentation was mounted 4.7 m above ground and the instruments used were a sonic anemometer (WindMaster, Gill Instruments, Lymington, UK) to measure the three-dimensional wind components and virtual (sonic) temperature, a LI-7700 open gas analyzer for CH_4 measurements (LI-COR Inc., Lincoln, NE, USA) and an LI-7500 open-path gas analyzer for CO_2 and H_2O measurements (LI-COR Inc., Lincoln, NE, USA). The tower was positioned on a very small island in the middle of the lake with the closest shore 1 km to the southeast (figure 1). The height of the tower base is approximately 20 cm above lake surface. The entire data set used here is for the open water seasons from September 2010 to August 2012.

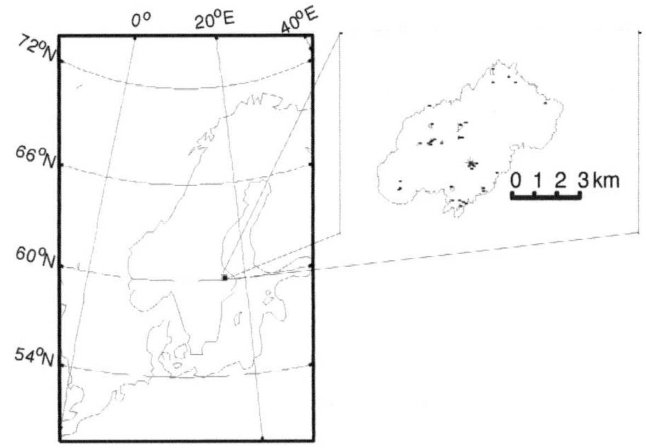

Figure 1. Left: Scandinavia and the location of the lake. Upper right: Lake Tämnaren (60°09'N, 17°20'E) with the tower location marked with a star.

2.2. EC data treatment

The high-frequency EC data were linearly de-trended and de-spiked over 30-min periods. Prior to flux calculations the sonic data were rotated using the double rotation technique. The time lag between the gas analyzers and the sonic was typically between 0.1 and 0.2 s and was calculated with the same procedure as in Sahlée *et al.* [3]. The gas analyzers measure the molar density fluctuations of the gas. The molar density of the gas can change either if the amount of molecules is changed or if the volume changes due to pressure, temperature, and humidity variations. To take into account concentration changes that are not real changes in the gas concentration the data were corrected according to Webb *et al.* [4] with some modifications for the LI-7700 analyzer [5]. For a more detailed description of the data analysis and quality we refer to previous studies using data from the same site [6, 7, 8].

3. Results

Analyses of the daily changes of CH_4 and CO_2 fluxes (FCH_4 and FCO_2) from Lake Tämnaren revealed that both FCH_4 and FCO_2 were substantially higher during nighttime compared to during daytime on average. This pattern was representative for a large part of the dataset (figure 2).

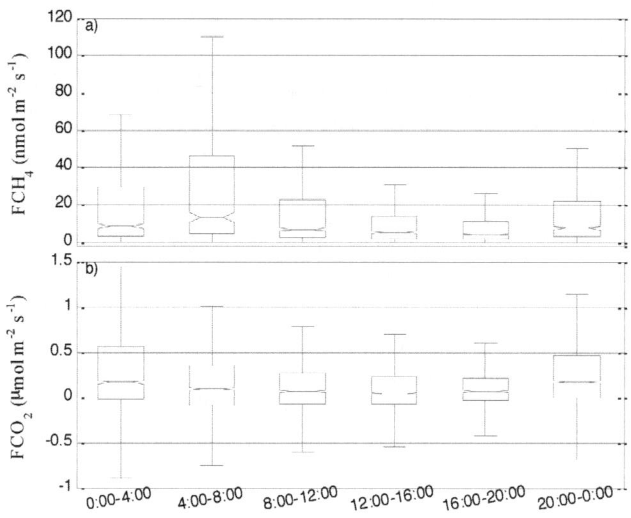

Figure 2. (a) FCH_4 and (b) FCO_2 as a function of time of day. The centerline of the notches in the boxes represent the median values, the edges of the boxes are the 25th (Q1) and 75th (Q3) percentiles, and the upper and lower whiskers represent Q3 + 1.5(Q3–Q1) and Q1–1.5(Q3–Q1), respectively.

Convective mixing in the water could be a possible mechanism enhancing the nighttime gas fluxes. Podgrajsek *et al.* [6] report that convection in the water might enhance FCH_4 through different mechanisms: enhancing the diffusive transport over the air-water interface, transporting CH_4-rich water from the bottom to the surface, and triggering ebullition events (bubbles with high concentration of methane).

The diffusive gas flux of CO_2 over the air-water surface is driven by the difference of partial pressure between the water and air (ΔpCO_2), and the efficiency of the gas transfer, the transfer velocity (k): $FCO_2 = K_o \cdot k \cdot \Delta pCO_2$, where K_o is a gas specific solubility constant. During one year the water CO_2 concentration was measured continuously (every 30 min) with a Submersible Autonomous Moored Instrument (SAMI) sensor (Sunburst Sensors, MT, USA). With these water measurements and air measurements of the CO_2 concentration and FCO_2, k_{meas} could be calculated. Normally k is parameterized with solely wind speed at 10 m. A commonly used parameterization of k for lakes is: $k_u = 2.07 + 0.215 \cdot u^{1.7}$ [9]. Podgrajsek *et al.* [8] observed that a vast amount of relatively large k_{meas} compared to k_u occurred when the water side velocity scale was large. Water side convection has previously been shown to increase the transfer velocity in marine environments [10]. Water side velocity scale is defined as $w_{w*} = (B \cdot h)^{1/3}$, where h is the mixed layer depth and B is the water side buoyancy flux defined as $B = \frac{gaQ_{eff}}{c_{pw}\rho_w}$, where g is the acceleration of gravity, a is the thermal expansion coefficient, Q_{eff} is the effective surface heat flux defined as the sum of the total heat flux, longwave radiation, and shortwave radiation, c_{pw} is the specific heat of water, and ϱ_w is the density of the water [11, 12] (figure 3).

It was concluded that the high nighttime FCO_2 fluxes could to a large degree be explained with enhanced diffusive transport over the air-water interface due to water side convection [8].

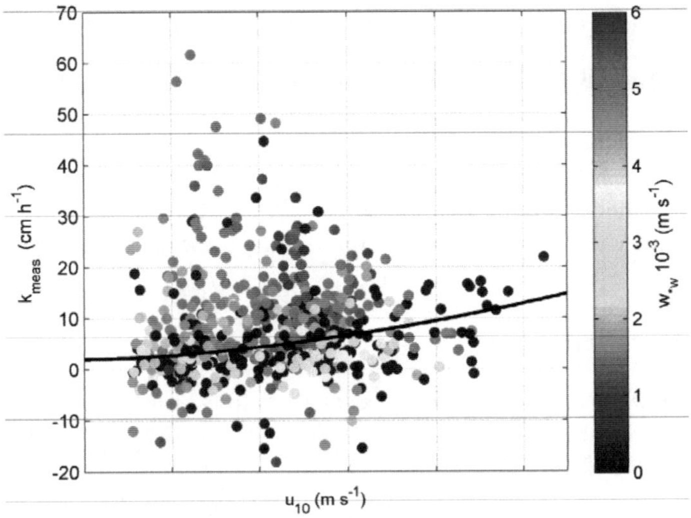

Figure 3. Transfer velocities for different wind speeds. The black line shows the parameterization from Cole and Caraco [9], k_u, and the color scale shows increasing water side velocity.

4. Summary and discussion

EC flux measurements of FCH_4 and FCO_2 from a lake in Sweden show that both gases have a diel cycle with higher values during nighttime compared to during daytime. The results show that water side convection can to a large extent explain the elevated fluxes during nighttime for both CH_4 and CO_2. This large variability in the fluxes measured during night and day and also between weeks and seasons, emphasizes the importance of making long-term continuous measurements when estimating total CH_4 and CO_2 fluxes from lakes. If gas flux measurements are not conducted during nighttime, potential high gas flux periods might be missed and estimations of the total amount of gas released from lakes to the atmosphere will be biased.

References

[1] Cole J J *et al* 2007 *Ecosystems* **10** 172
[2] Bastviken D *et al* 2011 *Science* **331** 50
[3] Sahlée E *et al* 2008 *Bound. Lay. Meteorol.* **126** 279
[4] Webb E K *et al* 1980 *Q. J. Roy. Meteorol. Soc.* **106** 85
[5] McDermitt D *et al* 2011 *Appl. Phys. B* **102** 391
[6] Podgrajsek E *et al* 2014 *J. Geophys. Res. Biogeosci.* **119** 236
[7] Sahlée E *et al* 2014 *Bound. Lay. Meteorol.* **150** 235
[8] Podgrajsek E *et al* 2015 *J. Geophys. Res. Biogeosci.* doi:10.1002/2014JG002781
[9] Cole J J and Caraco N F 1998 *Limnol. Oceanogr.* **43** 647
[10] Rutgersson A and Smedman A 2010 *J. Mar. Syst.* **80** 125
[11] Imberger J 1985 *Limnol. Oceanogr.* **30** 737
[12] Jeffery C D *et al* 2007 *Ocean Model.* **19** 161

Relationship between surface velocity divergence and turbulence microscale in open-channel flows with submerged strip roughness

T Okamoto, M Sanjou and I Nezu
Department of Civil and Earth Resources Engineering, Kyoto University, Kyoto, Japan

E-mail: takaaki.okamoto@water.kuciv.kyoto-u.ac.jp

Abstract. The present study investigates the effects of strip roughness on surface velocity divergence (SD) to develop physical modelling of gas transfer mechanisms in natural rivers. Particularly, turbulence measurements were conducted by PIV in a computer-controlled laboratory flume with varying water discharge and roughness spacing systematically, in order to obtain the space and time distributions of surface velocity divergence, turbulent kinetic energy and dissipation rate on the horizontal plane. Finally, a new empirical model for the surface velocity divergence was proposed considering turbulence microscales.

1. Introduction

McCready *et al.* [3] report that the surface velocity divergence (SD) is relevant to gas transfer processes at the air-water boundary. Therefore, the SD is highlighted in the gas transfer studies. The present study focuses on two-dimensional rough-bed flows, in which strip roughness elements are placed with constant spacing. It is well known that shedding vortices produced periodically behind the roughness element are transported toward the free surface. Although previous studies (Djenidi *et al.* [2], Pokarajac *et al.* [6]) point out that turbulence structure depends significantly on a ratio of roughness spacing to roughness height, many uncertainties remain about the relation between the SD and the coherent turbulence induced by the rough bed.

In this study we investigated hydrodynamic effects of strip roughness on the SD to develop physical modelling of gas transfer mechanisms in natural rivers. Particularly, turbulence measurements were conducted by PIV in a computer-controlled laboratory flume with varying water discharge, flow depth and roughness spacing systematically, in order to obtain the space and time distributions of SD, turbulent kinetic energy and dissipation rate. Finally, a new empirical model for the SD in natural rivers was proposed considering turbulence structure.

Content from this work may be used under the terms of the Creative Commons Attribution 3.0 licence. Any further distribution of this work must maintain attribution to the author(s) and the title of the work, journal citation and DOI.
Published under licence by IOP Publishing Ltd

Table 1. Hydraulic conditions

	ϕ	U_m(cm/s)	U_s(cm/s)	k_L(cm/s)	T_w(℃)	H (cm)	H/h
Case1-1	0.0	5.0	7.0	0.0013	24.6	12.0	8.0
Case1-2		10.0	13.3	0.0023	23.6		
Case1-3		20.0	24.3	0.0048	24.0		
Case2-1	3.0	5.0	7.4	0.0014	19.7		
Case2-2		10.0	14.8	0.0025	18.7		
Case2-3		20.0	27.4	0.0065	19.6		
Case3-1	5.0	5.0	7.1	0.0023	20.4		
Case3-2		10.0	13.2	0.0036	19.1		
Case3-3		20.0	27.4	0.0069	20.1		
Case4-1	8.0	5.0	6.0	0.0025	19.9		
Case4-2		10.0	14.2	0.0046	20.1		
Case4-3		20.0	32.6	0.0073	20.3		
Case5-1	12.0	5.0	7.2	0.0024	20.6		
Case5-2		10.0	14.9	0.0047	19.7		
Case5-3		20.0	30.2	0.0081	19.8		
Case6-1	16.0	5.0	7.3	0.0019	19.5		
Case6-2		10.0	13.8	0.0032	18.7		
Case6-3		20.0	30.8	0.0067	18.6		

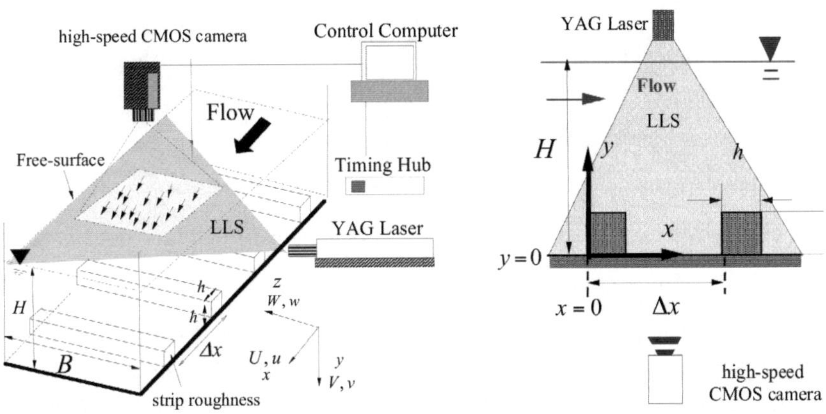

Figure 1. (left) Horizontal PIV system and (right) vertical PIV system

2. Measurement methods

Figure 1 shows the experimental set-up and the coordinate system. The experiments were conducted in a 16-m long and 40-cm wide glass flume. Streamwise, vertical and spanwise coordinates are x, y and z, respectively. The vertical origin, $y = 0$ was chosen as the channel bed. The time-averaged velocity components in each direction are defined as U, V and W, and the corresponding turbulent fluctuations are u, v and w, respectively. The strip roughness elements were placed on the flume bed with constant spacing in the streamwise direction. The roughness elements used were two-dimensional regular transverse square bars. The roughness height was $h = 15.0$ mm.

The measured region was located at about 7 m downstream from the channel entrance and the turbulent flow was fully developed. The 2.0-mm thick LLS (laser light sheet) was generated by a 3.0-W YAG laser using a cylindrical lens. The LLS was projected horizontally and vertically with a spatial resolution of about 0.29 mm per pixel. The LLS plane was illuminated together with tracer particles (diameter of 0.1 mm and specific density of 1.02) and captured by a high-speed CMOS camera. The instantaneous velocity vectors $(\tilde{u}, \tilde{w})$ were calculated by a PIV algorithm [5]. The PIV analysis was conducted by direct correlation, in which the interrogation window size was 25×25 pixels. When the correlation value between the first and second image patterns was less than 0.4, a local velocity vector was judged as an invalid vector, and interpolated velocity data were given to the corresponding position using surrounding valid vectors.

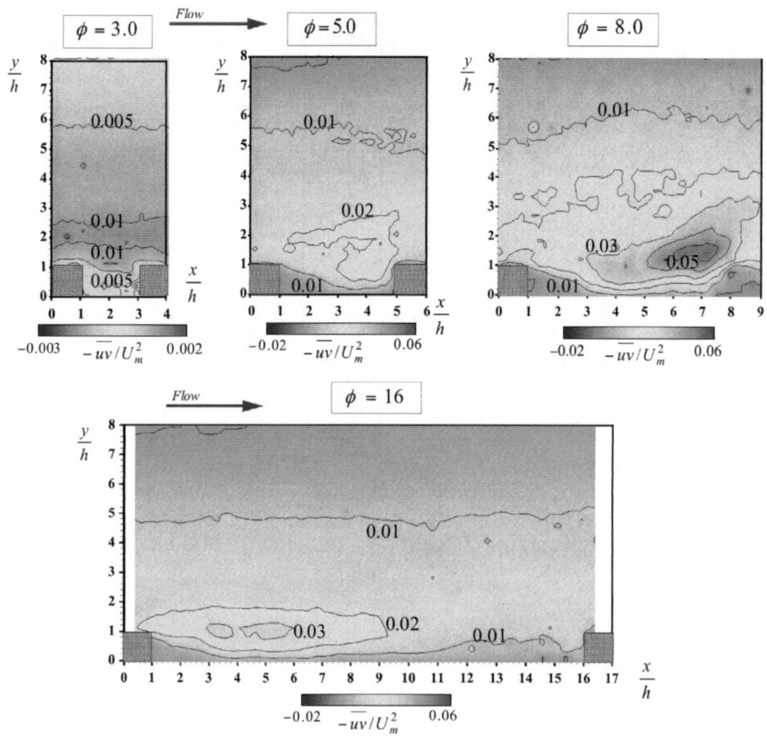

Figure 2. Contours of Reynolds stress by vertical PIV (U_m = 10 cm/s).

Table 1 shows the hydraulic conditions, in which U_m is the bulk mean velocity and U_s is the surface streamwise velocity in the centerline of the flume. $\phi \equiv \Delta x / h$ is a relative spacing of strip roughness elements, in which Δx is a streamwise distance between the neighboring roughness elements. The LLS positions are $y / H = 0.98$ for the horizontal PIV. The gas transfer velocity K_L was measured using two dissolved oxygen meters [4].

3. Results and discussion

3.1. Turbulence structure and surface divergence distribution for rough bed flow

Figure 2 shows the contours of the Reynolds stress $-\overline{uv}$ for $\phi = 3.0$, 5.0, 8.0 and 16.0. The values of $-\overline{uv}$ are normalized by the square of the bulk mean velocity U_m^2. Note that the Reynolds stress contours have the peak value at the canopy edge (*y*/*h*=1.0). As the coherent structure develops between elements, the peak values of $-\overline{uv}_{peak}(x)$ become larger downstream. It is also observed that for $\phi = 3.0$–8.0, the magnitude of the Reynolds stresses in the outer layer increases in proportion to the roughness spacing ϕ. In contrast, for $\phi = 16.0$, the Reynolds stress peak is located at *x*/*h* = 6.0. It is also observed that the Reynolds stress values become smaller as the flow approaches the next roughness element.

Figure 3 shows the contours of instantaneous surface velocity divergence $\tilde{\beta}$ and velocity vectors $(\tilde{u}, \tilde{w})$ on the horizontal plane (*y*/*H* = 0.98) in the cases of U_m = 10 cm/s for $\phi = 5.0$. The longitudinal positions of the roughness are indicated in figure 2. The velocity vectors are drawn in a movable coordinate subtracting the bulk-mean velocity U_m. $\tilde{\beta} < 0$ yields a convergence zone where momentum concentrates and is accompanied by a downward current. In contrast, $\tilde{\beta} > 0$ yields a divergence zone where upward current transferred from the flume bed results in momentum divergence.

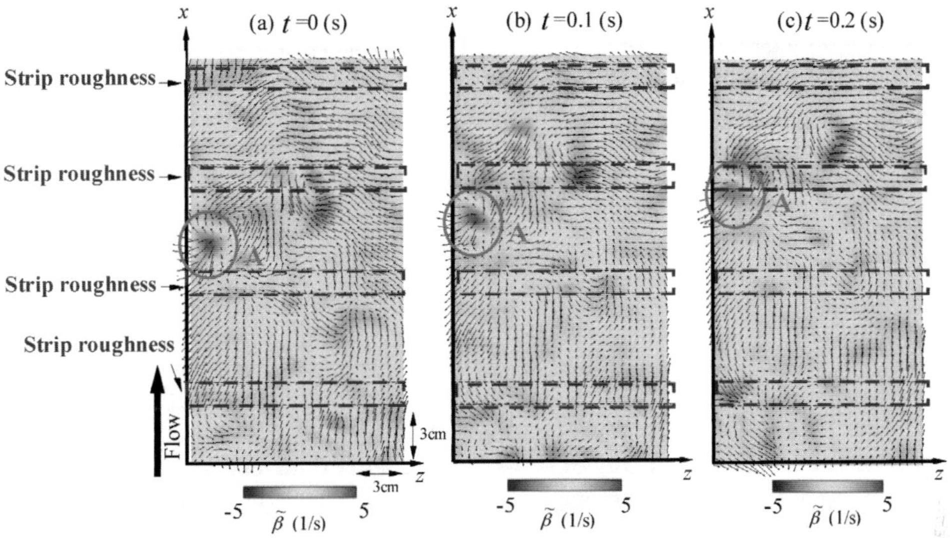

Figure 3. Time-series of the surface velocity divergence on the horizontal plane near the free surface (U_m= 10 cm/s, ϕ= 5).

$$\tilde{\beta} \equiv \frac{\partial \tilde{u}}{\partial x} + \frac{\partial \tilde{w}}{\partial z} = -\frac{\partial \tilde{v}}{\partial y} \quad (1)$$

From the contour of instantaneous surface divergence, the positive and negative divergence zones appear periodically (figure 3). In blue and red circles accompanied by large divergence values, we can distinguish the divergence and convergence zones of the free surface fluid by the direction of velocity vectors. The circle 'A' is transferred downstream, keeping its formation. Note that the convection velocity does not depend on the strip roughness position.

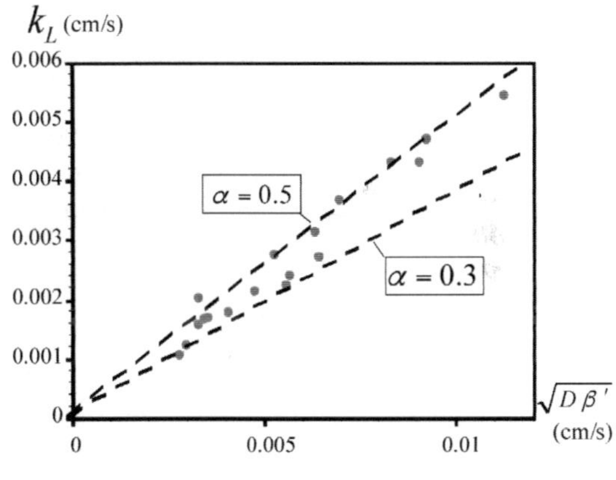

Figure 4. Relation between gas transfer velocity and the surface divergence intensity.

3.2. Effects of roughness spacing on surface divergence

Figure 4 shows the relationship between the gas transfer velocity k_L and the divergence intensity. McCready *et al.* [3] propose a surface divergence (SD) model as follows:

$$k_L = \alpha\sqrt{D\beta'} \quad (2)$$

in which D is molecular diffusivity of dissolved oxygen in water and $\beta' = \sqrt{\overline{(\tilde{\beta}-\overline{\beta})^2}}$ ($\overline{\beta}$: time-averaged surface velocity divergence) is the RMS value of the instantaneous SD, and α is a proportional coefficient. The linear relation is observed for flows with strip roughness and α ranges from 0.3 to 0.5.

Figure 5 shows the relationship between the roughness spacing and the divergence intensity. Note that a single peak appears at ϕ = 12, irrespective of the bulk-mean velocity U_m. Coleman *et al.* [1] report that the vortex strength formed over the roughness strip becomes larger when the relative spacing is

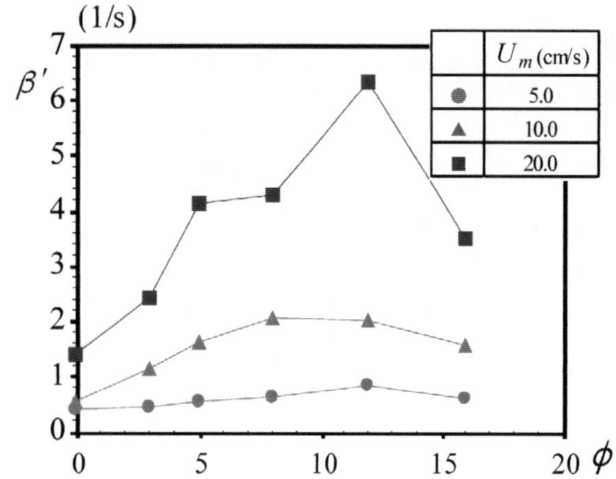

Figure 5. Relation between intensity of surface velocity divergence and relative spacing of strip roughness.

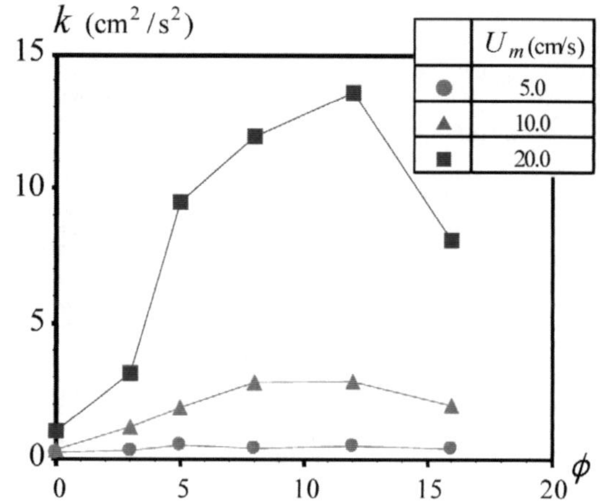

Figure 6. Relation between turbulent kinetic energy and relative spacing of strip roughness.

about $\phi = 8$ (see figure 2). It is thus inferred that the SD has a strong relation with the turbulent structures induced by rough-bed flow. Furthermore, the results revealed that the values of the SD become larger in high velocity conditions. This suggests that the Reynolds number may be a key factor on the production of the surface divergence.

It is generally difficult to measure the SD because we have to obtain the time-series of the velocity components $(\tilde{u}, \tilde{w})$ on the horizontal plane simultaneously. But the turbulence statistics, such as the turbulent kinetic energy and dissipation rate, can be analyzed using a single-point measurement method. Therefore, correlating the turbulence statistics and the surface divergence will contribute to the development of practical and useful prediction methods for the surface divergence and the related gas transfer velocity.

3.3. Turbulent kinetic energy

Figure 6 shows the relationship between the roughness spacing and the turbulent kinetic energy k. Turbulent kinetic energy k is defined as follows:

$$k = \frac{1}{2}(u'^2 + w'^2) \tag{3}$$

u' is the RMS value of the instantaneous streamwise velocity $\tilde{u}$, and w' is the RMS value of the instantaneous spanwise velocity $\tilde{w}$. The turbulent kinetic energy has a single peak at $\phi = 12$. This is in good agreement with figure 5 (the surface velocity divergence). Coleman *et al.* [1] conducted turbulence measurements and report that the Reynolds stress value reaches a maximum for $\phi = 8$. Our results are in good agreement. In cases of large relative spacing ($\phi = 16$), local turbulence structures are produced behind the roughness element, but they become smaller and disappear with distance from the roughness element (figure 2). In the cases of small roughness spacing, the spacing is smaller than the flow separation length and the turbulence structure does not develop between the roughness elements. Consequently, the mean flow and turbulence structures become uniform in the streamwise direction and near the wall boundary layer condition.

3.4. Physical modeling of SD with Kolmogorov scale

The turbulent dissipation rate ε is the transfer rate from turbulence energy toward heat per unit mass and time, which is often calculated by using Kolmogorov's scale.

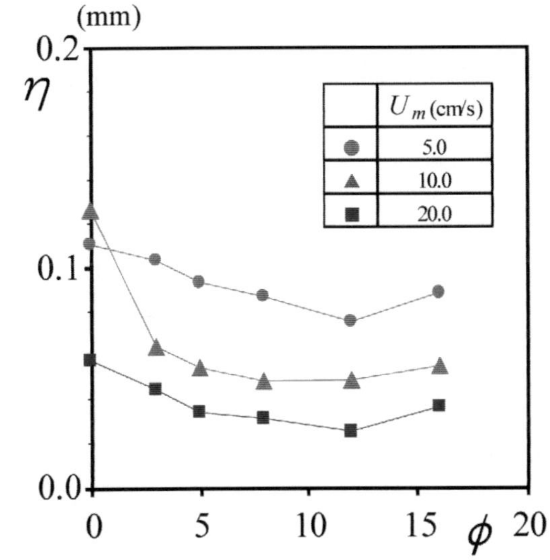

Figure 7. Relation between Kolmogorov length scale and relative spacing of strip roughness.

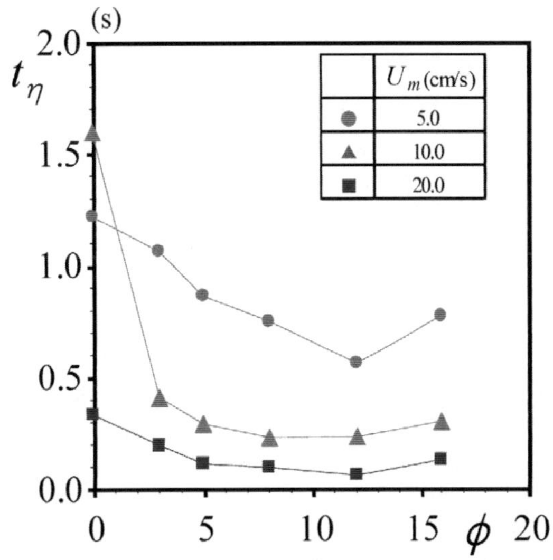

Figure 8. Relation between Kolmogorov time scale and relative spacing of strip roughness.

$$\frac{\varepsilon}{\nu} \equiv \frac{1}{2}\overline{\left(\frac{\partial u_i}{\partial x_j}+\frac{\partial u_j}{\partial x_i}\right)^2} \tag{4}$$

The present study evaluates it from the original definition because time series of velocity components could be obtained simultaneously for the multi-measured points. Here, we consider the dissipation rate on the free surface. The continuity equation leads to the following.

$$\frac{\partial v}{\partial y} = -\left(\frac{\partial u}{\partial x}+\frac{\partial w}{\partial z}\right) \tag{5}$$

Under the low Froude number condition, $v = 0$ and $\partial u/\partial y = \partial w/\partial y = 0$ can be assumed on the free surface. Combining equations (4) and (5) yields the following.

$$\frac{\varepsilon}{\nu} = 4\overline{\left(\frac{\partial u}{\partial x}\right)^2} + 4\overline{\left(\frac{\partial w}{\partial z}\right)^2} + 4\overline{\left(\frac{\partial u}{\partial x}\right)\left(\frac{\partial w}{\partial z}\right)} + \overline{\left(\frac{\partial u}{\partial z}+\frac{\partial w}{\partial x}\right)^2} \tag{6}$$

The right-hand side terms are calculated using measured two-dimensional PIV data.

Under Kolmogorov's local isotropy hypothesis, a probability density function of turbulence is defined by the viscous coefficient ν and turbulent energy dissipation ε. The characteristic length is obtained by these physical variables as indicated by equation (7). This corresponds to the length scale of minimal turbulent vortex. In the same way, a characteristic time scale could be defined by equation (8).

$$\eta \equiv \left(\frac{\nu^3}{\varepsilon}\right)^{1/4} \tag{7}$$

$$t_\eta \equiv \left(\frac{\nu}{\varepsilon}\right)^{1/2} \tag{8}$$

Figure 7 compares the Kolmogorov length among the different hydraulic conditions. The Kolmogorov length scale is smaller for a rough bed than for a smooth bed flow ($\phi = 0$). It reaches a minimum at $\phi = 12$ and increases slightly between $\phi = 12$ and 16. Figure 8 shows the relation between the roughness spacing and the Kolmogorov time scale. The same trend is observed as for the length scale (figure 7). These results suggest that both length and time scales become larger when turbulence production is comparatively small, and in contrast, they becomes smaller when greater turbulence is formed.

Figure 9 shows the relation between the time scale and the SD. A distinct linear relation is observed, as given by equation (9).

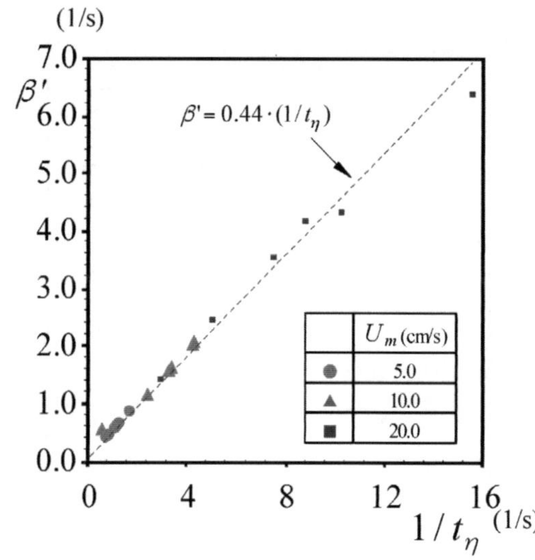

Figure 9. Linear relation between the Kolmogorov time scale and the surface divergence intensity.

$$\beta' = 0.44 \cdot (1/t_\eta) \tag{9}$$

The scatter of the measured data is small, making it more useful for the physical modelling.

4. Conclusions

We focused on the surface velocity divergence (SD) in open-channel flows with strip roughness. The horizontal PIV measurements obtained a time series of free-surface velocity divergence. Positive and negative divergence zones are transferred downstream with time in the same manner as the boil phenomena in a natural river surface.

The SD intensity relates with turbulence statistics, irrespective of the streamwise spacing of roughness elements. A distinct linear relation between the SD intensity and turbulence microscales, such as Kormogorov's scale, was observed for rough-bed flow.

References

[1] Coleman S *et al* 2007 *J. Engineering Mechanics* **133** 194
[2] Djenidi L *et al* 2008 *Exp. Fluid* **44** 37
[3] McCready M J *et al* 1986 *AIChE J.* **32** 1108
[4] Moog D B and Jirka G H 1999 *J. Hydraulic Eng.* **125** 3
[5] Nezu I and Sanjou M 2011 *J. Hydro-environment Res.* **5** 215
[6] Pokarajac D *et al* 2008 *Exp. Fluids* **45** 73

Relationship between surface velocity divergence and gas transfer in open-channel flows with submerged simulated vegetation

M Sanjou, T Okamoto and I Nezu

Department of Civil and Earth Resources Engineering, Kyoto University, Kyoto, Japan

E-mail: michio.sanjou@water.kuciv.kyoto-u.ac.jp

Abstract. Velocity and gas concentration measurements were carried out to reveal gas transfer phenomena in open-channel turbulent flows with flat bottom and submerged vegetation bottom conditions. A large-scale coherent vortex appears near the vegetation top due to shear instability, and the submerged vegetation was found to promote gas transfer beneath the air-water interface. Furthermore, we revealed a great dependency of gas transfer on vegetation density. The present measurement results propose a new surface divergence model with wide generality, connecting reasonably the gas transfer velocity and the surface divergence intensity in open-channel flows, irrespective of bottom roughness conditions.

1. Introduction

Surface velocity divergence is closely related to coherent motions induced by bottom turbulence. Instantaneous upward and downward motions occur together with the strong surface divergence, and it results in periodic renewal of the dissolved gas situation in the free surface. The surface velocity divergence plays a significant role in the gas transfer rate as pointed out by McCready *et al.* (1986), which introduces a theoretical relation, i.e., a surface divergence model (SD model).

It is generally well known that natural rivers are composed of various bottom roughness features such as submerged vegetation, gravel, sand waves, etc. They contribute to produce greater turbulence compared with flat bottom conditions. Practical prediction methods of the gas transfer rate are modelled by friction velocity. However, the friction velocity is hard to evaluate accurately in natural rivers. With development of video recording devices and particle image velocimetry (PIV) techniques, the horizontal velocity components could be obtained in the free surface, and it is possible that the SD model will be the most useful method to predict the gas transfer rate in rivers.

The present study focuses on the development of an SD-based model for open-channel flows with submerged plants. Surface divergence is found to be affected significantly by the free surface velocity. In the vegetated-bed cases, the divergence intensity is greater because of the shear vortex produced at the vegetation edge than the flat bottom condition. We propose a new practical model considering the turbulence dynamics. A proportionality coefficient of the original SD model depends on the water depth, and thus, a modified SD model was required that can consider the effects of water depth on gas transfer. We examined whether the free surface streamwise velocity and turbulent kinetic energy can

Content from this work may be used under the terms of the Creative Commons Attribution 3.0 licence. Any further distribution of this work must maintain attribution to the author(s) and the title of the work, journal citation and DOI.
Published under licence by IOP Publishing Ltd

be used as a proper velocity scale. In both cases of smooth and vegetated bed conditions, the modified SD model is proved to be useful without a dependency on the bulk mean velocity and depth.

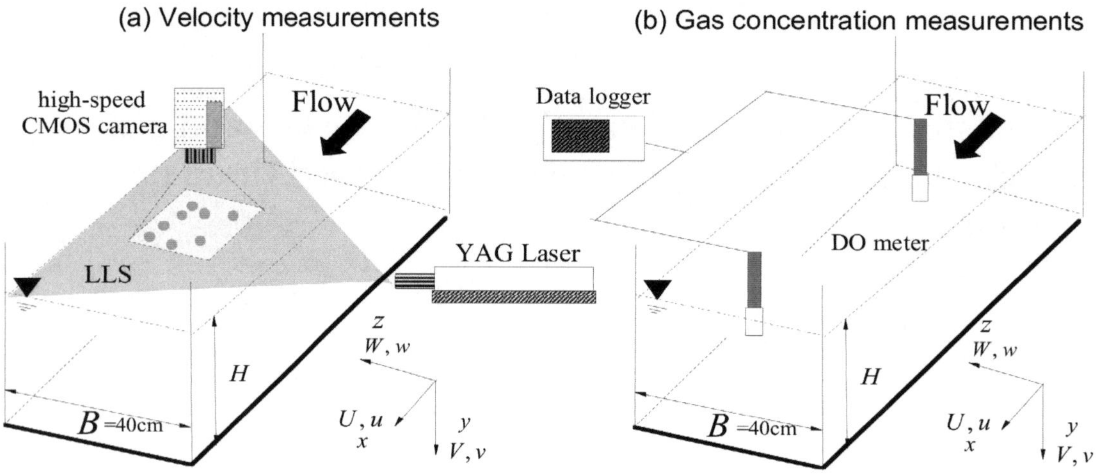

Figure 1. Experimental setup

2. Experimental procedure

Figure 1 shows a glass-sided water flume that is 16-m long, 40-cm wide, and 50-cm high, in which water current was generated by computer control and an electromagnetic flow meter. Streamwise, vertical and spanwise coordinates are x, y and z, respectively. The vertical origin, $y = 0$, was chosen as the free surface. The time-averaged velocity components in each direction are defined as U, V and W, and the corresponding turbulent fluctuations are u, v and w, respectively. The measured region was located at about 7 m downstream from the channel entrance, at which the turbulent flow was fully developed. A 2-W YAG laser (continuous wave) was used for the laser light sheet (LLS) as shown in figure 1a. The 2-mm thick LLS was projected horizontally in the surface elevation. The spatial resolution is about 0.22 mm per pixel. The LLS plane was illuminated together with tracer particles (diameter of 100 μm and density of 1.02 g/cm^3) and captured by a high-speed CMOS camera. The time variation of the instantaneous velocity vectors ($\tilde{u}$, $\tilde{w}$) was calculated by PIV algorithm (see Nezu and Sanjou 2011). The PIV analysis was conducted by direct correlation, in which the interrogation window size is 25×25 pixels. When the correlation value between the first and second image patterns is less than 0.4, a local velocity vector was judged an invalid vector, and an interpolated velocity value was given to the corresponding position using surrounding valid vectors.

Dissolved oxygen (DO) measurements were also conducted to evaluate directly the gas transfer velocity in the same manner as Moog *et al.* (1999). After calculation of the reaeration coefficient k_2, the gas transfer velocity k_L could be obtained multiplying k_2 and H. These measured transfer velocity values were modified to those expected at 20°C water temperature. It is well known that advection effects should be considered in the streamwise distance when we calculate k_L in unidirectional flows such as an open-channel stream. In the present experiment, two DO meters were aligned in the streamwise direction with 7-m span as shown in figure 1b. These measured signals were transferred to the data logger every 1.0 s.

Table 1. Hydraulic conditions for flat bottom flows.

H(cm)	U_m(cm/s)	U_s(cm/s)	Re	Fr	$\tan\theta$
8	5.0	6.2	4000	0.056	$1/10^4$
8	10.0	12.3	8000	0.113	$1/10^4$
8	20.0	24.6	16000	0.226	$1/10^4$
8	30.0	35.1	24000	0.339	$3/10^4$
8	40.0	45.6	32000	0.452	$5/10^4$
6	5.0	6.4	3000	0.065	$1/10^4$
6	10.0	12.5	6000	0.130	$1/10^4$
6	20.0	24.3	12000	0.261	$2/10^4$
6	30.0	36.0	18000	0.391	$4/10^4$
6	40.0	47.8	24000	0.522	$6/10^4$
4	5.0	6.3	2000	0.080	$1/10^4$
4	10.0	12.3	4000	0.160	$1/10^4$
4	12.5	15.1	5000	0.200	$1/10^4$
4	15.0	18.0	6000	0.240	$1/10^4$

Table 2. Hydraulic conditions for flows with submerged vegetation on the bottom.

λ	H(cm)	U_m(cm/s)	U_s(cm/s)	Re	Fr	$\tan\theta$
0	15	10.0	12.0	15000	0.082	$1/10^4$
0.0236	15	10.0	15.7	15000	0.082	
0.0947	15	10.0	16.8	15000	0.082	
0.3789	15	10.0	19.0	15000	0.082	

Table 1 shows hydraulic conditions, in which U_m is bulk-mean velocity and U_s is time-averaged free-surface streamwise velocity in the centerline of the flume. Subscript s means the free surface. Fr and Re are Froude number and Reynolds number defined using U_s and H, respectively. The DO measurements take long durations, from 10 to 40 hours, and thus, we kept constant the water and air temperatures by air conditioners. All fifteen hydraulic cases were chosen by varying U_s and H systematically. The aspect ratio of the cross-section, B/H, was 5.0 and larger, to avoid effects of secondary currents as pointed out by Nezu and Nakagawa (1993).

Table 2 indicates hydraulic conditions for vegetated open-channel cases, in which turbulent measurements were conducted in the same way as the smooth-bed open-channel flows. The water depth H is three times greater than the vegetation height h, i.e., $H = 3h = 15$ cm. The present study deals with only submerged vegetation flow, and focuses on the relation between the vegetation density and the gas transfer velocity. A vegetation model is a 5-cm height, 0.8-cm width, 1-mm thick plastic plate that cannot warp in the water stream. They are placed vertically on the base plate with a rectangular grid form. The submerged simulated element reduces the streamwise velocity within the canopy, and it results in shear instability accompanied by coherent vortices at the vegetation top. We conducted not only surface horizontal PIV but also vertical in the $x-y$ plane. The vegetation density has been defined in the following way.

$$\lambda \equiv nhb/S \tag{1}$$

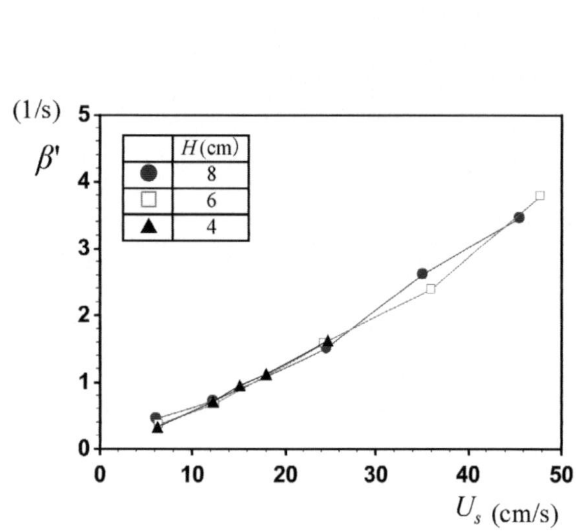

Figure 2. Relationship between surface velocity and surface divergence intensity.

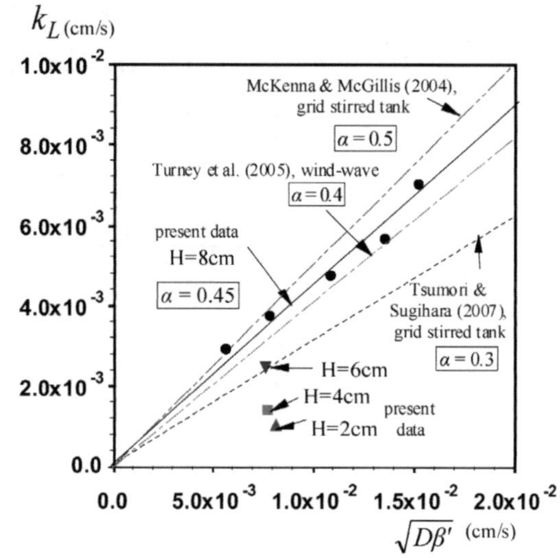

Figure 3. Examination of surface divergence model; comparison with previous measured data in grid stirred tank and wind waves.

S is the horizontal area corresponding to the vegetation zone and n is the total number of vegetation elements included in the vegetation zone. Grid span is 2.4 cm, 4.8 cm and 7 cm, corresponding to $\lambda = 03789, 0.0947$ and 0.0236, respectively.

3. Results in flat-bottom open-channel flows

The surface velocity divergence at the air-water interface $\tilde{\beta}$ could be defined as the following form, in which tilde represents an instantaneous component.

$$\tilde{\beta} \equiv \frac{\partial \tilde{u}}{\partial x} + \frac{\partial \tilde{w}}{\partial z} = -\frac{\partial \tilde{v}}{\partial y} \tag{2}$$

When $\tilde{\beta}$ is introduced instead of vertical velocity $\tilde{v}$, a transport equation of dissolved gas concentration is given by

$$\frac{\partial \tilde{c}}{\partial t} + \tilde{u}\frac{\partial \tilde{c}}{\partial x} + \tilde{w}\frac{\partial \tilde{c}}{\partial z} - \tilde{\beta} y \frac{\partial \tilde{c}}{\partial y} = D\frac{\partial^2 \tilde{c}}{\partial y^2} \tag{3}$$

This form implies that $\tilde{\beta}$ is an important factor when vertical gas transport is prevalent much more than longitudinal and spanwise ones. Figure 2 shows the relation between surface divergence intensity β' and the free-surface velocity or the water depth. β' is equal to RMS of the instantaneous surface velocity divergence. β' was calculated from time series of $\tilde{\beta}$ during 60 s. A fluctuation of water surface is negligibly small in the present hydraulic cases, and thus, a vertical velocity could be given by zero.

When we consider $\tilde{v}_s$ and v'_s situated in the free surface with distance, Δy, the following relation is obtained.

$$\beta' = v'_s / \Delta y \tag{4}$$

Turbulence intensities obey universal functions proposed by Nezu and Nakagawa (1993) except for near the bottom and free-surface regions, and they could be normalized by a bottom friction velocity

U_*. There remains uncertainty about turbulence structure near the surface; when we assume $v' \propto U_*$, a proportional relation $\beta' \propto U_*$ is obtained. In the open-channel flows, U_s is proportional to U_*, and it is then expected that β' is proportional to U_s. This tendency is observed in figure 2. A variation of β' is relatively small against the water depth.

In a surface renewal model (SR model), k_L is given by

$$k_L \propto \sqrt{Dr} \tag{5}$$

in which D is molecular diffusivity of dissolved oxygen in water and r is a surface renewal rate. Physical gas transfer modeling has been generally conducted on the basis of equation (5).

In contrast, McCready *et al.* (1986) proposed a surface divergence model (SD model) by connecting the gas transfer velocity and the surface velocity divergence.

$$k_L = \alpha\sqrt{D\beta'} \tag{6}$$

in which α is a proportional coefficient. The validity of the SD model was examined mainly for the grid-stirred tank experiments. However, not much is known about open-channel flows.

Figure 3 shows the relation between k_L and β', in which measured data in the grid-stirred tank and wind-induced water waves are indicated for the comparison. When the water depth is fixed, there is a linear relation in the present data as well as previous studies. α varies in a range of 0.3 to 0.5 in previous studies, and it is 0.45 in the present study of H = 8 cm. Therefore, the present proportional coefficient agrees well with those observed in other kinds of flow fields.

The linear relation is not observed among different water depth cases. Thus, a more practical SD model should be proposed for the open-channel cases such as natural rivers.

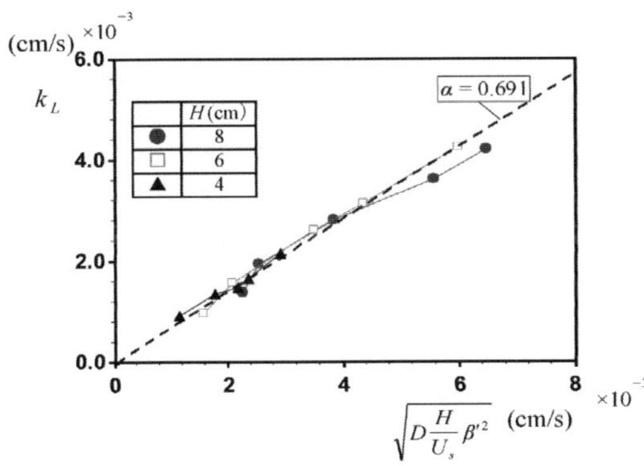

Figure 4. Relationship between gas transfer velocity and $\sqrt{DH\beta'^2/U_s}$

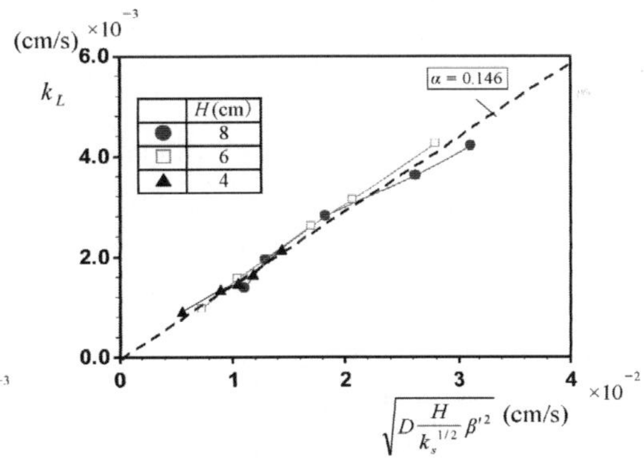

Figure 5. Relationship between gas transfer velocity and $\sqrt{DH\beta'^2/k_s^{1/2}}$

The proportional coefficient α in the original SD depends on the water depth. In a new proposed model, the water depth is chosen as characteristic length scale. It is further important in modeling of surface renewal rate to select only hydrodynamic values obtained by horizontal PIV measurements in field measurements in terms of practical use. When the surface streamwise velocity U_s is chosen as characteristic velocity, the new transfer model is indicated in the following way.

$$k_L \propto \sqrt{D\frac{H}{U_s}\beta'^2} \tag{7}$$

Figure 4 shows variation of the gas transfer velocity against $\sqrt{D\beta'^2 H/U_s}$. The linear relation could not be obtained with little dispersion, irrespective of the water depth.

The surface renewal process is correlated to the coherence turbulence, and thus, the turbulence statistics such as turbulent kinetic energy, turbulence intensity, and eddy advection velocity, etc., may be better for the characteristic velocity scale than the time-averaged velocity such as U_s. When we assume that vertical turbulence intensity at the free surface is ignorably small compared to streamwise and spanwise ones, the turbulent kinetic energy of the free surface k_s could be indicated by

$$k_s^{1/2} \equiv \frac{1}{2}(u'^2 + w'^2) \tag{8}$$

When k_s is used for U_s, the following relation could be introduced.

$$k_L \propto \sqrt{D\frac{H}{k_s^{1/2}}\beta'^2} \tag{9}$$

Figure 5 shows variation of the gas transfer velocity against $\sqrt{D\beta'^2 H / k_s^{1/2}}$, in which a linear relation is recognized. The standard deviation of $k_L / \sqrt{D\beta'^2 H / U_s}$ is greater than $k_L / \sqrt{D\beta'^2 H / k_s^{1/2}}$. This suggests that equation (9) is more reliable than equation (7). Then, equation (10) could be proposed for an experimental formula about gas transfer velocity in open-channel flows.

$$k_L / \sqrt{D\frac{H}{k_s^{1/2}}\beta'^2} = 0.146 \tag{10}$$

In the near future, we will examine equation (10) in a larger scale laboratory flume and natural rivers.

4. Gas transfer in vegetated-bottom conditions

Figure 6 shows the relation between the surface divergence intensity and the vegetation density. This result suggests that the divergence intensity increases with increased vegetation density. This may be because the large-scale coherent vortex plays a more significant role on the surface renewal in the denser vegetation case. Some previous studies offered that more remarkable shear instability could be observed in the denser vegetation, and corresponding integral scales of coherent structure increase significantly.

Figure 7 and figure 8 show significant relations of k_L and $\sqrt{DH\beta'^2 / U_s}$, and k_L and $\sqrt{DH\beta'^2 / k_s^{1/2}}$, respectively. In figure 7, the results of the non-vegetation case, i.e., only a base plate over the flume bed, agree with the linear relation of equation 6, while those in the vegetation cases shift below smooth bed results. This implies that U_s cannot be used as the velocity scale to consider large-scale turbulence induced by the coherent vortices near the vegetation top.

In figure 8, the results of vegetation cases are consistent with equation 10, irrespective of the vegetation density. This suggests that the prediction model using turbulent kinetic energy as the velocity scale has a wide generality.

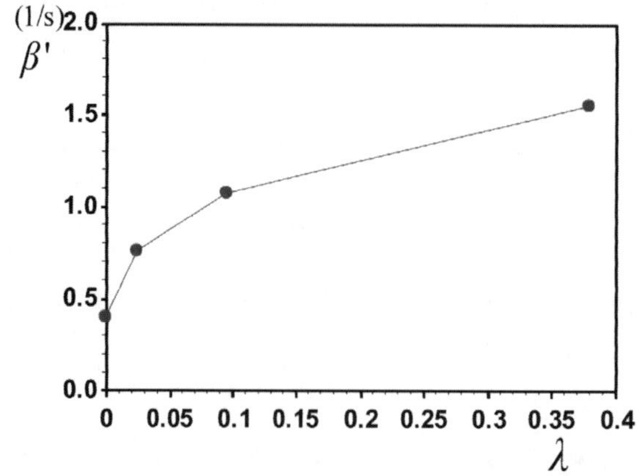

Figure 6. Relation of surface divergence intensity and vegetation density.

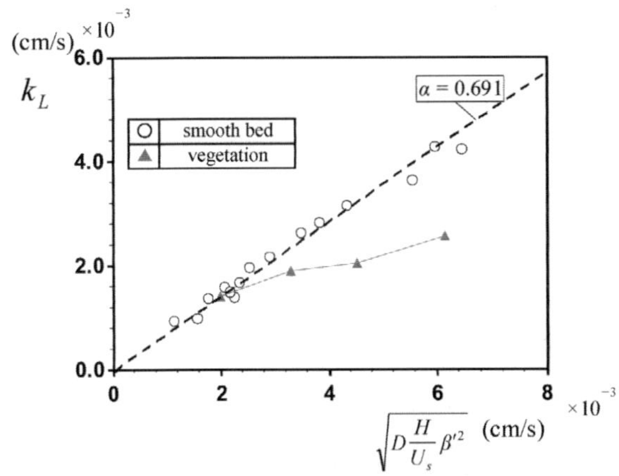

Figure 7. Relation between gas transfer velocity and $\sqrt{DH\beta'^2 / U_s}$ for both bottom types.

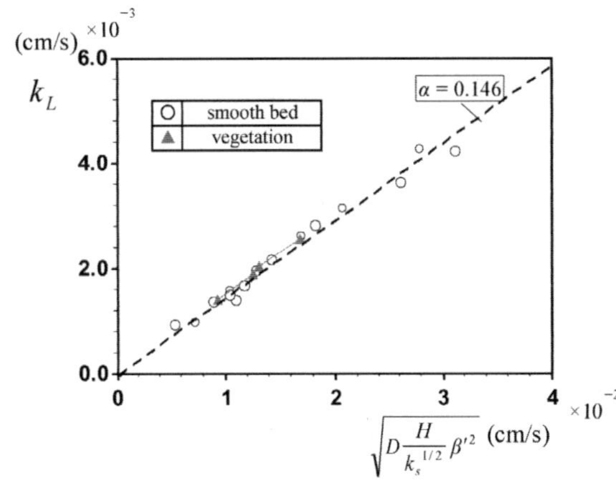

Figure 8. Relation between transfer velocity and $\sqrt{DH\beta'^2 / k_s^{1/2}}$ for both bottom types.

5. Conclusions

It was found that a proportionality coefficient of the original SD model depends on the water depth. The present study developed a modified SD model that considers the effects of water depth on gas transfer. We examined whether the free-surface streamwise velocity and turbulent kinetic energy are used as a proper velocity scale, and then, in both cases, the modified SD model is useful without a dependency on the bulk-mean velocity and the water depth in flat bottom conditions.

When the modified SD model using the free-surface velocity is applied to the vegetated open-channel flows, there is a large gap between the smooth bed case and the vegetated. In contrast, the modified model using the turbulent kinetic energy has a constant proportionality coefficient, irrespective of bottom roughness conditions. It is therefore indicated that the modified SD model, which gives the surface renewal rate using the turbulent kinetic energy and surface divergence, has the same proportionality coefficients for smooth and vegetated bottoms.

References

[1] McCready M J *et al* 1986 *AIChE J.* **32** 1108
[2] Nezu I and Sanjou M 2011 *J. Hydro-environment Res.* **5** 215
[3] Moog D B and Jirka G H 1999 *J. Hydraulic Eng.* **125** 3
[4] Nezu I and Nakagawa H 1993 *Turbulence in Open-Channel Flows* (CRC Press)
[5] Komori S *et al* 1989 *J. Fluid Mech.* **203** 103

IOP Conf. Series: Earth and Environmental Science **35** (2016) 012017 doi:10.1088/1755-1315/35/1/012017

Quantifying air-sea gas exchange using noble gases in a coastal upwelling zone

C C Manning[1,2], R H R Stanley[3], D P Nicholson[2] and M E Squibb[4]

[1] Massachusetts Institute of Technology/Woods Hole Oceanographic Institution Joint Program in Oceanography/Applied Ocean Science and Engineering, Woods Hole, MA 02543, USA
[2] Department of Marine Chemistry and Geochemistry, Woods Hole Oceanographic Institution, Woods Hole, MA 02543, USA
[3] Department of Chemistry, Wellesley College, Wellesley, MA 02481, USA
[4] Center for Ocean Solutions, Stanford University, Stanford, CA 94305, USA

E-mail: cmanning@whoi.edu

Abstract. The diffusive and bubble-mediated components of air-sea gas exchange can be quantified separately using time-series measurements of a suite of dissolved inert gases. We have evaluated the performance of four published air-sea gas exchange parameterizations using a five-day time-series of dissolved He, Ne, Ar, Kr, and Xe concentration in Monterey Bay, CA. We constructed a vertical model including surface air-sea gas exchange and vertical diffusion. Diffusivity was measured throughout the cruise from profiles of turbulent microstructure. We corrected the mixed layer gas concentrations for an upwelling event that occurred partway through the cruise. All tested parameterizations gave similar results for Ar, Kr, and Xe; their air-sea fluxes were dominated by diffusive gas exchange during our study. For He and Ne, which are less soluble, and therefore more sensitive to differences in the treatment of bubble-mediated exchange, the parameterizations gave widely different results with respect to the net gas exchange flux and the bubble flux. This study demonstrates the value of using a suite of inert gases, especially the lower solubility ones, to parameterize air-sea gas exchange.

1. Introduction

Noble gases dissolved in seawater are biologically and chemically inert, making them excellent tracers of numerous physical processes that control gas saturation states in the ocean (e.g., bubble-mediated and diffusive gas exchange, temperature change, atmospheric pressure change, ice melting, diapycnal mixing, deepwater formation and ventilation) [1–3]. By simultaneously measuring several inert gases with a range of physical properties (e.g., diffusivity, solubility, and dependence of solubility on temperature), we can separately quantify many of these physical processes [4, 5]. For example, the solubility of the noble gases decreases with decreasing atomic mass, and the lower solubility noble gases (He and Ne) have a larger portion of their total gas exchange flux driven by bubbles. The dependence of bubble flux on solubility can be explained as follows: for lower solubility gases the atmospheric concentration is high relative to the water concentration, and thus when air bubbles dissolve in the water, the bubbles will generate a larger percent increase in the gas concentration, compared to a higher solubility gas. Additionally, the saturation state of the higher solubility (heavier) noble gases changes more dramatically in response to surface heating/cooling, and mixing of water masses with different temperatures, due to the stronger dependence of the solubility on temperature compared to the lower solubility (lighter) noble gases. Parameterizations of physical processes from inert gases can be applied to bioactive gases, to obtain more accurate estimates of processes including biological productivity [3, 6] and denitrification [7, 8].

Here, we use a quasi-Lagrangian time-series of the five stable noble gases (He, Ne, Ar, Kr, and Xe) in Monterey Bay, CA, along with a vertical model, to evaluate the performance of four gas exchange

Content from this work may be used under the terms of the Creative Commons Attribution 3.0 licence. Any further distribution of this work must maintain attribution to the author(s) and the title of the work, journal citation and DOI.
Published under licence by IOP Publishing Ltd

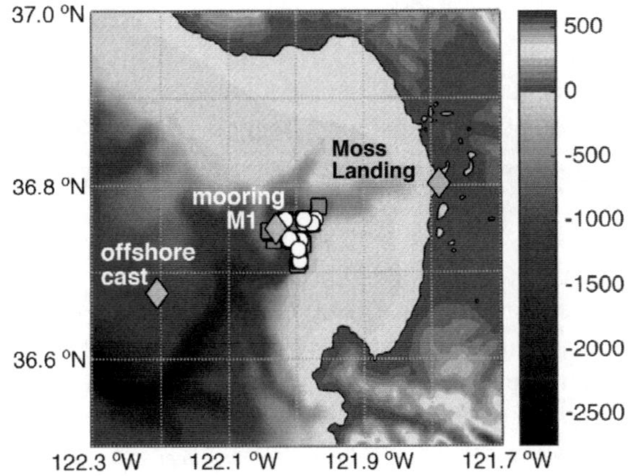

Figure 1. Map of the study site. Pink diamonds indicate the locations of Moss Landing, CA, mooring M1, and the offshore cast. Grey squares are locations of CTD casts and white circles are locations where surface samples were collected (sometimes colocated with CTD casts, sometimes in between casts). Color bar shows elevation and bathymetry in meters.

parameterizations that differ in their treatment of diffusive and bubble-mediated air-sea gas exchange. These data are some of the few published measurements of a suite of noble gases in a coastal upwelling zone, where environmental factors such as increased levels of surfactants and reduced fetch for wind wave generation may cause the relationship between wind speed and gas exchange rates to differ from the relationship in the open ocean [9, 10]. In a forthcoming publication, we will apply the noble gas-based parameterizations of physical processes to O_2 concentration and isotope measurements, in order to quantify net community production and gross primary production during the time-series.

2. Methods

2.1. Cruise description

We participated in a seven-day quasi-Lagrangian cruise in Monterey Bay, CA, USA, on the R/V *Western Flyer*, Sept 27–Oct 3, 2014. During the cruise, rosette casts to $\sim$180 m were conducted four times per day at roughly 06:00, 12:00, 18:00, and 00:00 local time. Immediately prior to nearly every cast, vertical profiles of microscale turbulence to $\sim$70 m were obtained using a vertical microstructure profiler (VMP-200, Rockland Scientific). Rates of turbulent kinetic energy dissipation were calculated using two perpendicular air-foil type shear probes on the VMP-200. The Nasmyth spectrum was used to recursively estimate dissipation and diapycnal diffusivity (K_z) [11]. Roughly three high-quality turbulence profiles were obtained every 6 hr.

All CTD casts were performed and all noble gas samples were collected in the vicinity of mooring M1 (21 km west of Moss Landing, CA), which is maintained by the Monterey Bay Aquarium Research Institute (figure 1). Starting on Sept 28, 12:35 and throughout the rest of the cruise, we selected where to collect CTD and turbulence profiles by following an autonomous underwater vehicle (AUV) drifting at $\sigma_t \approx 25.2$ kg m^{-3} (25–45 m depth), near the base of the thermocline. Thus, the cruise was quasi-Lagrangian, but with respect to the base of the thermocline, rather than the mixed layer. We also collected noble gas samples from one offshore cast (37 km west of Moss Landing, CA), which indicated that there is minimal spatial variability in noble gas distributions, with respect to both saturation anomaly and concentration.

2.2. Noble gas data

In this paper, we present noble gas data (He, Ne, Ar, Kr, and Xe concentrations) from discrete samples, which were collected from the ship's underway seawater line and from Niskin bottles. The samples were collected in copper tubes, sealed with a cold pressure welder, and extracted in the Isotope Geochemistry Facility at Woods Hole Oceanographic Institution [12]. Noble gas abundances were measured on a pulse counting quadrupole mass spectrometer using the system described in Stanley *et al.* (2009) [13]. The dissolved concentration of each gas (in $\mathrm{cm}^3_{STP}\ \mathrm{g}^{-1}$ or mol kg^{-1}) was determined, using selected ion monitoring [13] and including isotope dilution analysis for Kr and Xe. The estimated error (combined precision and accuracy) is $\pm$0.27, 0.27, 0.24, 0.25, and 0.27 %RSD (relative standard deviation) for He, Ne, Ar, Kr, and Xe, respectively. As we did not analyze samples in replicate from this dataset, the errors are based on a different dataset of duplicate samples that were collected and analyzed by the same methods immediately following this dataset. Three samples with evidence of air contamination (based on anomalies in the relative excess concentration of the five noble gases in the sample, compared to the other samples at a similar depth) were eliminated from the dataset.

Because the rate of gas exchange is dependent on the deviation of the gas concentration from equilibrium, accurate solubility functions are critical for accurate parameterization of air-sea gas exchange and other physical processes. Hamme and Emerson (2004) have published high-quality solubility data for Ne and Ar (errors of $\pm$0.30 and 0.13%, respectively) based on equilibration of water with air at a range of temperatures and salinities [14]. For the other gases in seawater, researchers have typically used the solubility data of Weiss (1971) for He, Weiss and Kyser (1978) for Kr, and Wood and Caputi (1966) for Xe [15–17]. Errors of $\sim$2% may be present in these solubility functions because they were determined for 1 atm of each pure gas, and these data must be extrapolated over several orders of magnitude to calculate the solubility with respect to air. For example, Weiss (1971) determined the solubility of Ne in addition to He, and his reported seawater Ne solubility was $\sim$1.5% lower than Hamme and Emerson (2004) [14, 15]. Additionally, the published Kr and Xe solubilities in seawater are subject to uncertainties in the atmospheric mole fraction of each gas [18]. Therefore, for He, Kr, and Xe, we use currently unpublished solubility data from Lott and Jenkins (personal communication, 2015), who determined the solubility of He, Kr, and Xe from 0 to 36.6 PSS, and from $<$1 to 35 °C by equilibration of water with air. An additional advantage of the Lott and Jenkins dataset is that the solubilities were all measured on the same samples, using the same instrument. Thus the He, Kr, and Xe solubilities of Lott and Jenkins should be more consistent internally compared to data compiled from three different papers (two different laboratories). At the typical sea surface conditions during our study (S = 34.4 PSS, T = 16 °C), the solubilities of Lott and Jenkins are 2.2% greater for He, 1.3% greater for Kr, and 0.1% less for Xe, compared to the published solubilities.

In the main paper, we show the data and model results calculated with the Hamme and Emerson (2004) solubilities for Ne and Ar [14], and the Lott and Jenkins solubilities for He, Kr, and Xe. In the Supporting Information, we show the results using the published solubilities for He, Kr, and Xe, and we provide the noble gas concentration data along with ancillary data (salinity, temperature, atmospheric pressure, etc.) so that it can be used by future investigators with the most accurate solubility functions available. The choice of solubility functions, however, does not affect our overall conclusions; all models simulate the He data better when the Lott and Jenkins solubility is used instead of Weiss (1971) [15].

For the gas molecular diffusivity, which also enters into the gas exchange parameterizations, we used the freshwater data of Jähne (1987) [19] adjusted by –0.138% per ppt salinity (–4.75% adjustment for 34.4 PSS, the typical salinity in our study) for He, Ne, Kr, and Xe, and extrapolated values for Ar [20].

3. Model description

3.1. Data input and model setup

We developed a 1D vertical model for the time-series including diffusion and gas exchange for each noble gas. We ran the model using four different gas exchange parameterizations and evaluated the ability of each parameterization to accurately simulate the mixed layer saturation anomalies and concentrations measured throughout the cruise. The model was initialized with CTD profiles of temperature and salinity, and idealized noble gas profiles, described in more detail below. The model run proceeded as follows: at each time step (20 s), for each gas, the air-sea gas exchange flux was calculated and then the gas concentration in the surface box (upper 1 m) was adjusted, based on this flux. Then the diffusive flux at the edge of each box (1 m spacing) was calculated, which mixed the change in gas concentration from gas exchange throughout the mixed layer and deeper into the water column. We used a Crank–Nicolson diffusion scheme to calculate the flux between each box, based on the vertical diffusivity rates measured every 6 hr. The vertical diffusion rate was variable with depth and with time, and the model did not explicitly prescribe a mixed layer depth. Measured diffusivity was generally orders of magnitude higher near the surface where gases are mixed much more vigorously, compared to deeper waters below the mixed layer. Our vertical diffusivity determined from microstructure profiles is more accurate compared to other published studies where the diffusivity was estimated for the base of the mixed layer only.

We ran the model from Sept 28, 00:00–Oct 3, 02:00 local time. We linearly interpolated all data to the model time step of 20 s. CTD profiles of temperature and salinity were binned to the depth grid spacing of 1 m, from 0.5–69.5 m, and then linearly interpolated between casts. Temperature and salinity profiles from each CTD downcast (every 6 hr) were used in the model for calculating the simulated gas saturation anomalies. The turbulence profiles measured prior to each CTD cast (roughly three turbulence profiles per CTD cast) were averaged, and then linearly filtered to 1 m depth spacing (1–70 m) before interpolation to the model time step.

Measurements of wind speed, atmospheric pressure, and relative humidity were obtained at 10-min intervals from sensors on mooring M1 (figure 1). There were two wind sensors on the mooring: a Vaisala ultrasonic anemometer and a RM Young propeller anemometer. For this study we used data from the ultrasonic anemometer because it was specified to have higher accuracy and a lower detection limit. The wind sensors were both at 4 m height above the sea surface and the wind speeds were extrapolated to u_{10}, the wind speed at 10 m height, using the equation $u_b = u_a(b/a)^{0.11}$ with a and b the sensor and extrapolated heights, respectively [21]. During the time period included in the model (excluding the upwelling event, discussed below), the root mean square deviation (RMSD) of the two sensors was 0.38 m s^{-1} (9.8%). The mean difference between the two sensors was 0.29 m s^{-1}, with the anemometer giving higher wind speed 86% of the time. The model results were very similar regardless of which anemometer dataset was used. For example, the net gas exchange flux typically differed by 13% or less when the model was run with data from either anemometer.

The gas saturation anomaly was used in all the gas exchange parameterizations, and is defined in Δ notation as

$$\Delta C = (C_w/C_{eq} - 1)100\% \tag{1}$$

with C_w and C_{eq} the water and saturation equilibrium gas concentrations, respectively. The equilibrium concentration of each gas at each time step was calculated with respect to local sea level pressure and humidity, specifically

$$C_{eq} = C_{eq,ref} \times (P_{slp} - P_{H2O})/(P_{ref} - P_{H2O,eq}) \tag{2}$$

where $C_{eq,ref}$ is the reference equilibrium concentration at the measured salinity and temperature for 1 atm total pressure of air with water vapor at 100% humidity. The pressure terms are all expressed in atm: P_{ref} = 1 atm is the total reference pressure of air, P_{slp} is the local sea level pressure, and P_{H2O} and $P_{H2O,eq}$ are the water vapor pressures in situ and at equilibrium (100% relative humidity), respectively, calculated from the salinity, temperature, and relative humidity [22].

The model was initialized with an idealized profile for each gas (figure 2), because we did not have a depth profile at the start of the cruise. The concentration was fit to be similar to the other profiles at depth. The initial surface concentration of each gas was determined by fitting the model to either the concentration or saturation anomaly of the first sample, which was collected on Sept 28, 06:15. The fit was determined by minimizing the RMSD between the sample and model results for the four parameterizations. Fitting the models to the initial sample concentration versus saturation anomaly gave slightly different results because the saturation anomaly of each discrete sample was calculated based on the temperature and salinity of the water sampled, rather than the modeled temperature and salinity, which was interpolated from the CTD downcasts. There were sometimes differences between the real-time underway and interpolated CTD profiles (for underway samples) and offsets in the temperature and salinity profiles between the upcast and downcast (for Niskin samples). Therefore, the model and sample equilibrium concentrations were sometimes different. Below, we separately report the performance of the models optimized to concentration and saturation anomaly (section 4).

3.2. *Adjustment for upwelling*

We observed evidence for upwelling of colder water with higher noble gas and lower O_2 concentrations between the casts on Sept 29, 18:30 and Sept 30, 06:16 (figure 3). Temperature sensors at mooring M1 recorded a >1 °C decrease in temperature at 1, 10, and 20 m depth between Sept 29, 19:40 and Sept 30, 06:10. This time period coincided with the strongest winds and the strongest northerly (upwelling-favorable) winds observed during the time-series (winds were 10.0–11.8 m s^{-1} from 18:40–21:30). Additionally, diapycnal diffusivity was enhanced during the upwelling event, from roughly 10–20 m on Sept 29, 23:30 and then from 30–40 m on Sept 30, 05:30, suggesting vertical propagation of wind-driven turbulence from the surface to deeper waters (figure 3).

Our diffusion-based model was not able to reproduce the upwelling event. Therefore, we manually reset the gas profiles at Sept 30, 06:16, the time of the first cast after upwelling. The saturation anomaly of each gas throughout the upper 25 m was reset to the same value for all models, and

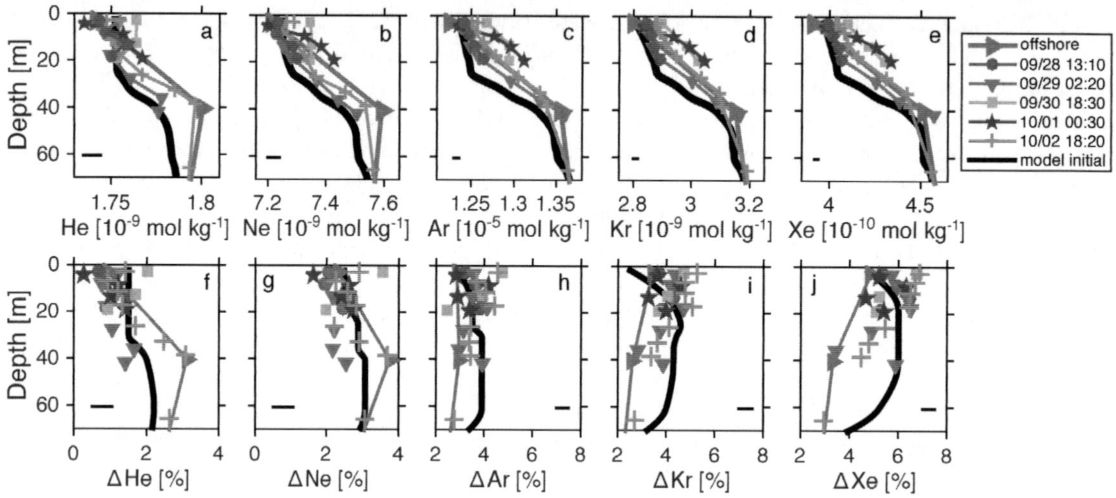

Figure 2. Measured profiles and idealized initial profile of noble gas concentrations (a–e) and saturation anomalies (f–j) during the cruise. The offshore cast was collected on Sept 30, 21:30 and its location is shown in figure 1. The black line is the idealized profile that was used to initialize the model on Sept 28, 00:00. The horizontal black lines show the estimated measurement error.

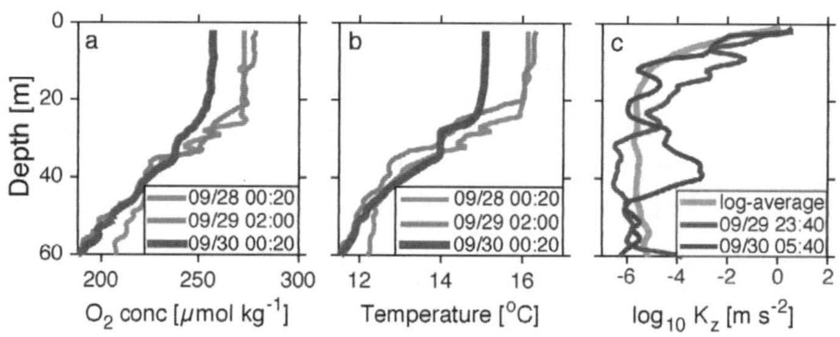

Figure 3. a) O_2, b) temperature, and c) diapycnal diffusivity (K_z) profiles collected during the cruise. a–b) The profiles on Sept 30, 00:20, during the upwelling, showing a decrease in mixed layer temperature and O_2. c) Diffusivity in the upper 40 m was enhanced following the upwelling, compared to the log-average (geometric mean).

these saturation anomalies were determined by minimizing the RMSD of the model results for the four different parameterizations to the first two samples collected after the upwelling event, on Sept 30, 06:33 and 07:10. Gas concentrations below 25 m were not adjusted because the temperature and O_2 profiles below 25 m remained similar following the upwelling event (figure 3). We fit the model to either the mean concentration or mean saturation anomaly of these two samples. This approach was identical to the approach used to determine the initial surface concentrations. We excluded samples collected during the upwelling event (Sept 29, 19:33 and Sept 30, 01:30) when assessing the model performance.

We did not add a continuous upwelling flux throughout the cruise (e.g., using a published upwelling index for the region). Upwelling events are episodic in this region during the fall [23], and our temperature and mooring data indicated that only one upwelling event occurred during our time-series, overnight Sept 29–30. Additionally, we were able to simulate the remainder of the time-series without a continuous upwelling flux.

We omitted advection from the mass balance because we did not have sufficient spatial coverage to resolve advective fluxes. This omission is likely not a serious limitation because our measurements at an offshore station suggested that the dissolved gas concentrations and saturation anomalies were very similar offshore to near M1 (figure 2). Furthermore, we evaluate the performance of the models using only He and Ne, which should be the least affected by advective fluxes and mixing between different water masses. These gases will have the lowest lateral variability in concentration and saturation anomaly, due to the very small temperature dependence of their solubilities (0.2 and 0.7% $°C^{-1}$ for He and Ne, respectively) [2, 5, 24]. Additionally, our data show that the surface He concentration was barely altered by the upwelling event, a clearly non-Lagrangian process.

3.3. Choice of gas exchange parameterizations

We ran the model using four different gas exchange parameterizations: Liang *et al.* 2013 (L13), Nicholson *et al.* 2011 (N11), Stanley *et al.* 2009 (S09), and Sweeney *et al.* 2007 (Sw07) [2, 3, 25, 26]. N11 and S09 were developed from inverse models of noble gas data, with S09 using a three-year time-series of monthly profiles of He, Ne, Ar, Kr, and Xe concentration at the Bermuda Atlantic Time-series Study (BATS) site in tandem with a 1D oceanic mixed layer model, and N11 using a global dataset of Ne, Ar, N_2/Ar, and Kr/Ar measurements in tandem with an ocean general circulation model. L13 used a large eddy simulation (LES) model coupled to a bubble population model to parameterize the bubble-mediated exchange and did not use oceanic gas measurements to tune the model; however, the L13 parameterization reproduced oceanic data well. With all three parameterizations, the total gas exchange flux is calculated as the sum of diffusive and bubble-mediated exchange:

$$F_t = F_d + F_b \tag{3}$$

with F_t the total flux, F_d the diffusive flux, and F_b the bubble flux. Each parameterization further separates the bubble-mediated flux into two components:

$$F_b = F_c + F_p \tag{4}$$

where F_c is the complete trapping bubble flux from typically smaller bubbles that completely dissolve and F_p is the partial trapping bubble flux from typically larger bubbles that partially dissolve. The exact equations for F_d, F_c, and F_p differ between authors, as we discuss below.

Finally, Sw07 used a global dataset of dissolved inorganic radiocarbon data to quantify the rate of uptake of anthropogenic (bomb) radiocarbon into the ocean. Sw07 does not explicitly include the bubble-mediated component of air-sea gas exchange, i.e., it includes F_d but not F_b. The Sw07 parameterization is within the range of other gas exchange parameterizations derived from oceanic measurements that do not include a separate term for bubble-mediated exchange [27–30], and is used as F_d in N11.

All of the parameterizations use u_{10} as the only environmental forcing variable. In L13, u_{10} is converted to u_* using an empirical formula for the drag coefficient [31, 32]. Therefore, none of the models explicitly include variables such as surfactants, precipitation, and/or fetch.

S09, Sw07, and N11 all parameterize F_d using an equation of the form

$$F_d = A u_{10}^2 (Sc/660)^{-0.5} (C_{eq} - C_w). \tag{5}$$

Here A is an empirical constant and Sc is the Schmidt number of the gas at the water temperature and salinity. Sc is defined as the ratio of the kinematic viscosity of seawater to the gas diffusivity. N11 and Sw07 both use A = 0.27 cm hr^{-1}, whereas S09 uses A = 0.30 cm hr^{-1}, and will thus yield a diffusive flux that is 11% greater for a given wind speed and gas concentration. The comparison to F_d calculated from L13 is more complex because u_* is a nonlinear function of u_{10}. At u_{10} = 5 m s^{-1}, F_d of L13 is approximately equal to F_d of S09 and N11. At lower wind speeds, F_d of L13 exceeds S09 and N11 (by a factor of five at u_{10} = 1 m s^{-1}), and at higher wind speeds F_d of L13 is less than S09 and N11 (by a factor of two at u_{10} = 10 m s^{-1}). The differences in F_d for L13 (scaled to u_*) versus the other parameterizations (scaled to u_{10}) also have a slight dependence on the gas Schmidt number.

For the bubble-mediated fluxes F_c and F_p, the differences between parameterizations are not simple to characterize because they are functions of additional variables such as the gas saturation anomaly, diffusivity, solubility, and atmospheric mole fraction, as well as u_{10}. The dependence on these factors can differ between models. For example, F_p is scaled to the gas diffusivity, D, as $F_p \propto D^{0.5}$ in N11 and $F_p \propto D^{2/3}$ in S09. L13 sets $F_p \propto (Sc/660)^{-2/3}$. All three parameterizations assume that F_c is independent of D.

We report the differences between parameterizations for F_d, F_c, and F_p in our model below. We refer the reader to the authors' original papers for more details on each parameterization [2, 3, 25, 26]. In the Supporting Information, we provide MATLAB functions for calculating gas exchange fluxes using each of the four tested parameterizations, as well as several additional parameterizations that only include F_d [20].

4. Results

We evaluated the performance of each parameterization by comparing the model results to near-surface measurements (2–4 m depth, 15 samples total). We show the gas time-series with respect to saturation anomaly as well as with respect to concentration (figure 4), and we calculated the error using both parameters. For the most soluble gases, Ar, Kr, and Xe, all models gave very similar results and simulated the surface measurements well. The four models diverged for He and Ne, which have higher diffusivity and lower solubility, making them more sensitive to differences in the treatment of bubble-mediated gas exchange.

The concentration plots for Ar, Kr, and Xe clearly show the impact of the upwelling event overnight on Sept 29–30: near-surface gas concentrations following the upwelling were 2–4% higher than before the upwelling (figure 4c–e). The concentration increase for Ar, Kr, and Xe following the upwelling was driven by the replacement of warmer water by colder water with higher equilibrium gas concentrations (but similar saturation anomalies for each gas). In contrast, the concentrations of He and Ne changed by <1 % due to the upwelling because these gases' solubility has a lower dependence on temperature and therefore the equilibrium gas concentration was similar before and after upwelling. Additionally, because the upwelling event coincided with high winds, some of the concentration increase for He and Ne during this period can be attributed to bubble-driven supersaturation rather than temperature change.

There was one sample following the upwelling, shaded grey in figure 4, that fell below the model results for Ar, Kr, and Xe concentration, but not saturation anomaly. This sample was collected in warmer water (16.4 °C) compared to the four surface samples taken within 24 hr of that sample (15.2–15.7 °C) and therefore it had a lower equilibrium gas concentration (C_{eq}). It is possible that this sample may have been from a different water mass compared to the rest of the time-series, but it was included in all of our error estimates because we had no specific reason to discount it.

For each gas, the initial surface concentration and surface concentration after upwelling were set by fitting the models to the measured concentrations of the first sample and the first two samples after upwelling, respectively. Due to the measurement uncertainty for these initialization values, we performed a Monte Carlo error analysis where the initial concentration and concentration after upwelling were simultaneously varied randomly 100 times with a normal distribution, with the optimized concentrations (figure 4) set as the mean and the measurement error set as the standard deviation. We evaluated the performance of the different parameterizations by calculating the RMSD between all measured near-surface sample concentrations and the coincident modeled concentrations, for each noble gas and each parameterization. We then repeated the error analysis, but instead fit the models to the measured saturation anomaly of the first sample, and the first two samples after upwelling, and calculated the RMSD with respect to saturation anomaly. We report the results for each parameterization as the mean ±

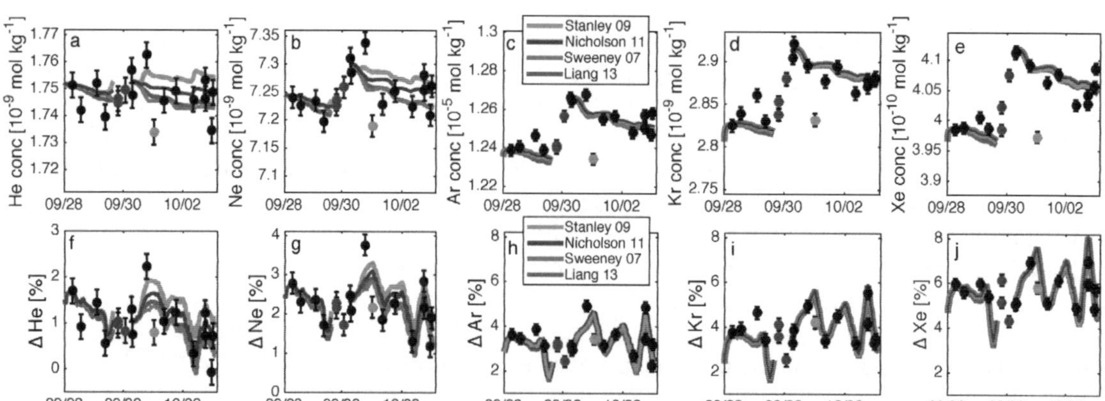

Figure 4. Near-surface gas concentrations (a–e) and saturation anomalies (f–j) from samples and model results. The gap in the models around midnight Sept 30 corresponds to the upwelling event. Error bars reflect one standard deviation error in concentration measurement and do not include solubility uncertainty. Maroon circles are samples during the upwelling event that were not included in the Monte Carlo error analysis. The grey circle is an unexplained outlier that was included in the error analysis.

IOP Conf. Series: Earth and Environmental Science **35** (2016) 012017 doi:10.1088/1755-1315/35/1/012017

Table 1. Performance of the four parameterizations in simulating the surface He and Ne data, with and without adjustment for upwelling. RMSD is reported as the mean ± standard deviation, from the Monte Carlo error analysis. The % best fit indicates the frequency that a specific parameterization gave the best fit (lowest RMSD) of the four parameterizations.

He, with upwelling adjustment

Parameterization	He concentration RMSD (10^{-9} mol kg^{-1})	He saturation RMSD (%)	% best fit, He concentration	% best fit, He saturation
Sweeney 2007	8.3 ± 0.8	0.47 ± 0.05	8	3
Nicholson 2011	8.5 ± 1.1	0.46 ± 0.05	25	30
Stanley 2009	10.4 ± 1.5	0.56 ± 0.07	2	3
Liang 2013	7.9 ± 0.7	0.44 ± 0.04	65	64

Ne, with upwelling adjustment

Parameterization	Ne concentration RMSD (10^{-8} mol kg^{-1})	Ne saturation RMSD (%)	% best fit, Ne concentration	% best fit, Ne saturation
Sweeney 2007	3.8 ± 0.4	0.48 ± 0.05	2	1
Nicholson 2011	3.8 ± 0.4	0.48 ± 0.06	32	33
Stanley 2009	4.2 ± 0.6	0.55 ± 0.08	10	8
Liang 2013	3.7 ± 0.3	0.46 ± 0.04	56	58

He, without upwelling adjustment

Parameterization	He concentration RMSD (10^{-9} mol kg^{-1})	He saturation RMSD (%)	% best fit, He concentration	% best fit, He saturation
Sweeney 2007	9.6 ± 0.3	0.53 ± 0.02	0	0
Nicholson 2011	7.8 ± 0.6	0.43 ± 0.04	45	36
Stanley 2009	13.7 ± 0.8	0.77 ± 0.08	0	0
Liang 2013	7.7 ± 0.2	0.42 ± 0.01	55	64

Ne, without upwelling adjustment

Parameterization	Ne concentration RMSD (10^{-8} mol kg^{-1})	Ne saturation RMSD (%)	% best fit, Ne concentration	% best fit, Ne saturation
Sweeney 2007	7.6 ± 0.4	0.96 ± 0.05	0	0
Nicholson 2011	4.7 ± 0.3	0.58 ± 0.04	0	0
Stanley 2009	3.4 ± 0.1	0.44 ± 0.02	100	100
Liang 2013	6.8 ± 0.4	0.85 ± 0.05	0	0

standard deviation of the RMSD determined from the Monte Carlo error analysis for both concentration and saturation anomaly (table 1).

Unfortunately, the fact that the highest wind speeds in our time-series coincided with the upwelling event made it more difficult to discriminate between parameterizations. If the models are run through the full cruise duration, without resetting the concentrations after the upwelling, the four gas exchange parameterizations predict substantially different trajectories for He and Ne during and following the high wind event (figure 5). As wind speed increases, the magnitudes of F_d, F_p, and F_c increase and so do the differences between models. Below, we discuss the advantages and disadvantages of including upwelling versus not including upwelling in our model, and the insights gained from both approaches.

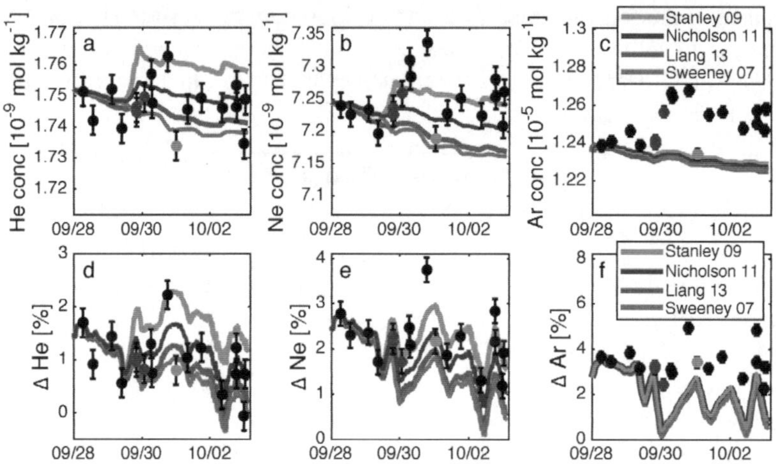

Figure 5. Model results without upwelling adjustment. Near-surface gas concentrations (a–c) and saturation anomalies (d–f) from samples and model results that were optimized to give the best fit to the first sample. All plotted samples were included in the error analysis.

5. Discussion

5.1. Modeled surface concentrations and saturation anomalies

When the model is adjusted for upwelling, the L13 model has the most skill in simulating the surface He and Ne data, yielding the lowest RMSDs of the four parameterizations (table 1). L13 is the most accurate parameterization for predicting surface He concentration and saturation 65% and 64% of the time, respectively. Additionally, L13 is the most accurate parameterization for predicting surface Ne concentration and saturation 56% and 58% of the time, respectively. N11 is most accurate 25–33% of the time for He and Ne concentration and saturation. The S09 and Sw07 models were each the most accurate parameterization in 10% or less of the Monte Carlo simulations for He and Ne. In general, the RMSD for He and Ne concentration and saturation is similar for L13, N11, and Sw07, and higher for S09. The modeled concentrations and saturation anomalies for He and Ne, when including upwelling, looked very similar for L13 and Sw07, although as shown in table 1, L13 was usually the most accurate parameterization. However, a large part of the apparent skill of Sw07 at reproducing the observed He data comes from the upwelling adjustment. The gas added by upwelling compensates for the lack of gas added by an explicit bubble-generated flux.

For He, when the model is run without adjustment for upwelling, the differences between parameterizations, and the need for an explicit bubble-generated flux to produce the observed He supersaturation become clearer (figure 5). Without the upwelling adjustment, the Sw07 parameterization predicts lower He concentrations and saturation anomalies than the other parameterizations, especially toward the end of the time-series. If the time-series were extended, the He concentration in Sw07 would continue to decay toward equilibrium (He_{eq}), and Sw07 would therefore underestimate the true He concentration. While we do not have a long enough time-series to say for certain, the fact that all but one surface He sample in our time-series are supersaturated suggests that a model that predicts equilibrium concentrations at steady state, such as Sw07, would be incorrect. An explicit bubble flux into the ocean is needed to generate consistently supersaturated surface waters for gases that are insensitive to temperature change. The solubility of He only changes by 0.2% $^{\circ}C^{-1}$.

Without the upwelling adjustment, L13 is most accurate for predicting surface He concentration and saturation anomaly 55% and 64% of the time, and N11 is most accurate 45% and 36% of the time, respectively. The performance of L13 and N11 for He is similar with and without the upwelling adjustment. However, notably, without the upwelling adjustment, the S09 parameterization appears to overestimate the bubble flux for He during the high winds that coincided with the upwelling.

For Ne, when the model is run without an adjustment for upwelling (figure 5), S09 is the most accurate parameterization for Ne concentration and saturation anomaly 100% of the time and has the lowest RMSDs of the four parameterizations, even though S09 has the highest RMSDs for Ne when the upwelling adjustment is included, and the highest RMSDs for He without the upwelling adjustment. This result suggests that Ne is an intermediate case, where upwelling, bubble-mediated fluxes, and diffusive fluxes are all important in setting the near-surface gas concentration. In the model runs without upwelling, S09 predicts high bubble fluxes that increase the Ne concentration to be similar to the true concentration after upwelling. The case with upwelling is likely more reasonable for Ne, although it does unfortunately overestimate the skill of the Sw07 model, due to the short duration of the time-series. Again, if the time-series were extended, the Ne concentration modeled with Sw07 would decay toward equilibrium, which is not consistent with the observations. All of the surface Ne samples are supersaturated, despite the solubility of Ne having a week dependence on temperature (0.7% $^{\circ}C^{-1}$), which supports the need for an explicit bubble-mediated flux into the ocean to generate some portion of the observed supersaturation.

For Ar, Kr, and Xe, all parameterizations simulate the data well after the correction for upwelling. For these gases, the RMSD of each parameterization is very similar and we conclude that all parameterizations have similar skill in simulating the heavier noble gases for our dataset. With a somewhat longer time-series, and/or without the upwelling event, the differences between the parameterizations would likely be clearer, and we might be able to see the importance of bubble-mediated exchange for somewhat more soluble gases, such as Ar. When the model is run without adjustment for upwelling, all four parameterizations predict similar trajectories for Ar, Kr, and Xe and underestimate the concentrations and saturation anomalies of these gases after the upwelling event (figure 5).

These results demonstrate the different factors that controlled the change in concentration for each gas overnight on Sept 29–30. For He, bubble-mediated exchange generated a small concentration increase (<1%). For Ne, upwelling and bubble-mediated exchange were both important in producing a moderate concentration increase (~1%). For Ar, Kr, and Xe, upwelling/mixing caused a large (2–3%) concentration increase. Because Ar, Kr, and Xe were barely affected by the high wind event that coincided with the upwelling, their concentrations could not be used to constrain the proportion of the surface water that was replaced with colder water with higher gas concentrations.

Over the whole time-series, the surface concentrations of Ar, Kr, and Xe were primarily controlled by upwelling and diffusive gas exchange. These more soluble gases contrast with He, which was primarily controlled by bubble-mediated and diffusive gas exchange, and Ne, which was controlled by a combination of upwelling, bubble-mediated gas exchange, and diffusive gas exchange.

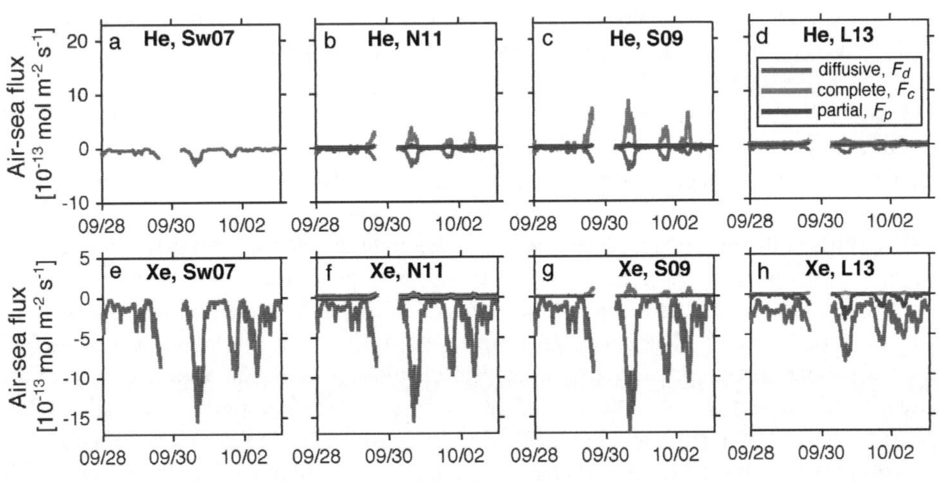

Figure 6. Air-sea gas fluxes of He (a–d) and Xe (e–h) during the time-series including adjustment for upwelling and modeled using all four parameterizations. Positive fluxes are into the ocean. The flux is separated into three components: diffusive flux (F_d), complete bubble trapping (F_c), and partial bubble trapping (F_p). The Sw07 parameterization only includes F_d.

IOP Conf. Series: Earth and Environmental Science **35** (2016) 012017 doi:10.1088/1755-1315/35/1/012017

Table 2. Comparison of air-sea fluxes, based on data plotted in figure 6 (i.e., including adjustment for upwelling). The F_d and F_b are mean values, averaged over the whole time-series. Positive fluxes are into the ocean. Liang 2013 predicts $F_p < 0$ and $F_c > 0$ for He and Xe. Therefore, the fraction of F_b from F_p is negative for He ($F_b > 0$) and over 100% for Xe ($F_b < 0$).

Parameterization	He F_d	He F_b	He fraction of F_b from F_p	Xe F_d	Xe F_b	Xe fraction of F_b from F_p
	[10^{-13} mol m^{-2} s^{-1}]		[%]	[10^{-13} mol m^{-13} s^{-1}]		[%]
Sweeney 2007	–4.7	0	–	–2.7	0	–
Nicholson 2011	–5.8	3.7	11	–2.7	0.1	41
Stanley 2009	–7.5	7.9	3	–3.0	0.1	1
Liang 2013	–5.0	0.6	–16	–2.8	–0.4	103

5.2. Modeled air-sea fluxes

For each gas, the differences between parameterizations, or lack thereof, can be understood by examining the gas fluxes predicted by each parameterization (figure 6 and table 2). The parameterizations differ greatly in their prediction of the total bubble flux, and the fraction of the total bubble flux from partial trapping versus complete trapping. The Sw07 and N11 parameterizations use the same equation for F_d, but Sw07 does not include the bubble terms F_c and F_p. The magnitude of F_d for He is 19% less in Sw07 compared to N11 because there is no bubble-mediated exchange to increase the saturation anomaly, and thus the diffusive flux of He. The magnitude of the He bubble flux, F_b, is greatest for S09 (double N11 and twelve times L13); thus S09 predicts the highest He concentrations throughout the time-series.

The lower skill of the S09 parameterization compared to N11 and L13 in simulating He (both with and without the upwelling adjustment), and the improved performance of S09 in simulating Ne when the upwelling adjustment is not performed, both suggest that S09 may be overestimating bubble-mediated exchange in this environment. One potential explanation is that the S09 parameterization was tuned using He solubility data that was ∼1% too low (saturation anomalies that were ∼1% too high), compared to the LJ15 solubility used in this model. Assuming that the S09 model was fit to reproduce He saturation anomalies that were too high, it may overestimate bubble fluxes. As we observe, S09 does predict the largest bubble-mediated fluxes of the four parameterizations. When we run the model using the He solubility of Weiss (1971) [15], which is ∼2% lower than the He solubility of LJ15, all models underestimate the near-surface He concentrations and saturation anomalies (figures 2 and 3 in Supporting Information). The S09 parameterization comes closest to simulating the He data when using the Weiss (1971) solubility [15] because it predicts the largest bubble fluxes.

The N11 and L13 parameterizations were fit to inert gas data not including He, and the L13 parameterization was not tuned using any oceanic gas data. Therefore, N11 and L13 would not have errors related to uncertainties in He solubility. This result underscores the need for accurate gas solubility functions in order to realistically interpret oceanic gas data [14].

The modeled air-sea fluxes help to explain why the parameterizations all give very similar trajectories for Xe, Kr, and Ar in figure 4. For Xe, the Sw07 and N11 models predict nearly the same F_d because the diffusive flux greatly exceeds the bubble flux (table 2). The absolute and relative values of F_c and F_p for Xe vary widely between the different parameterizations. L13 is unique among the parameterizations in simulating a bubble stripping effect in which heating-induced supersaturation exceeds the bubble overpressure effect, resulting in a net removal of gas by partial bubble trapping (F_p). The S09 and N11 parameterizations always give net bubble flux into the ocean for F_p, even for supersaturated gases. N11, S09, and Sw07 predict similar F_d for Xe and have a small or zero bubble flux (the magnitude of F_b is 5% or less of the magnitude of F_t). For L13, the sum of F_d and F_b for Xe gives a net flux that is very similar to F_d for the other parameterizations.

The choice of optimizing the initial conditions to concentration versus saturation anomaly does not significantly affect our interpretations (table 1). For example, the modeled total air-sea flux (F_t) for each gas varies by 7% or less between the two optimization values. Additionally, the Monte Carlo error analysis results are very similar when the wind speed from the propeller anemometer is used rather than the sonic anemometer. The modeled F_t for each parameterization is generally within 13% for either anemometer, and both wind speed sensors predict L13 is most accurate 64–77% of the time for He and 56–72% of the time for Ne.

6. Conclusions and future work

This study is complementary to others that have demonstrated the value of using oceanic inert gas measurements in tandem with models to quantify air-sea gas exchange fluxes [2, 3, 6]. We demonstrated that short-term, high-frequency measurements of inert gases and diapycnal diffusivity can be used to quantify air-sea gas exchange in coastal regions. We found that accurately parameterizing bubble-mediated exchange was necessary to simulate the near-surface measurements of He and Ne collected during our time-series. The tested parameterizations gave a wide range of results for the direction and magnitude of the net bubble flux, and the proportion of partial versus complete bubble trapping, indicating that there are still large uncertainties in models of bubble-mediated gas exchange. Higher wind speed conditions, and/or a longer period of Lagrangian observations would have resulted in a greater divergence between the parameterizations for He and Ne.

The parameterizations that displayed the most skill in simulating the He observations were Liang 2013 [25] followed by Nicholson 2011 [2]. This result was observed regardless of whether we ran the model with or without an adjustment for upwelling. For Ne, the parameterization of L13 was most accurate when the model was adjusted for upwelling partway through the cruise. For Ar, Kr, and Xe, all parameterizations gave very similar results and simulated the observations well, after adjustment for upwelling. For these heavier gases and the moderate wind speeds observed during our study, we conclude that diffusive exchange driven by temperature change was more important than bubble-mediated fluxes in controlling the gas concentrations of Ar, Kr, and Xe. Due to the complication of upwelling during our cruise, a longer uninterrupted Lagrangian time-series and/or higher average wind speeds would have enabled clearer discrimination between parameterizations for the heavier gases, and an evaluation of the importance of bubble-mediated exchange for Ar.

The four tested parameterizations performed fairly well, despite not explicitly incorporating factors other than wind speed that may affect gas exchange rates in the coastal ocean (e.g., fetch and surfactant concentration). Evaluating these and other gas exchange parameterizations against in situ measurements in a wide range of environmental conditions, and for longer periods of time, is an important goal of future work.

Acknowledgments

We thank Captain Andrew McKee of the R/V *Western Flyer* and all cruise participants (crew and science team) for a successful cruise. Mooring data were collected and made available by the Biological Oceanography Group (PI: Francisco Chavez) at the Monterey Bay Aquarium Research Institute. Noble gas sample extraction, analysis, and data reduction was performed by Kevin Cahill, Bill Jenkins, and Dempsey Lott of the Woods Hole Oceanographic Institution Isotope Geochemistry Facility. We also thank one anonymous reviewer for their thoughtful comments that improved the manuscript. Funding for this work was provided by NSF award OCE-1060840 to RHR Stanley, NSF award OCE-1129644 to DP Nicholson, an Ocean Ventures Fund award from the WHOI Academic Programs Office to CC Manning, and graduate scholarships from NSERC and CMOS to CC Manning.

Supporting Information

The Supporting Information for this paper includes the noble gas data presented in the paper and plots of the model results using published solubility functions for He, Kr, and Xe, and a toolbox of MATLAB functions for calculating gas exchange fluxes with the parameterizations cited in this paper. The gas_toolbox is also available at http://dx.doi.org/10.5281/zenodo.46569 [20]. Use of the gas_toolbox also requires installation of the Gibbs-Seawater Oceanographic Toolbox for MATLAB (http://www.teos-10.org/software.htm) [33].

References

[1] Loose B and Jenkins W J 2014 *Geophys. Res. Lett.* **41** 2835–2841
[2] Nicholson D P, Emerson S R, Khatiwala S and Hamme R C 2011 *Proceedings of the 6th International Symposium on Gas Transfer at Water Surfaces, Kyoto University Press, Kyoto, Japan*

[3] Stanley R H R, Jenkins W J, Lott D E and Doney S C 2009 *J. Geophys. Res. Oceans* **114** C11020
[4] Hamme R C and Severinghaus J P 2007 *Deep Sea Res. I* **54** 939–950
[5] Hamme R C and Emerson S R 2002 *Geophys. Res. Lett.* **29** 2120
[6] Hamme R C and Emerson S R 2006 *J. Mar. Res.* **64** 73–95
[7] DeVries T, Deutsch C, Primeau F, Chang B and Devol A 2012 *Nat. Geosci.* **5** 547–550
[8] Chang B X, Devol A H and Emerson S R 2012 *Global Biogeochem. Cy.* **26**
[9] Upstill-Goddard R C 2006 *Estuar. Coast. Shelf S.* **70** 388–404
[10] Frew N 1997 *The Sea Surface and Global Change* ed Liss P and Duce R (Cambridge Univ. Press, New York) pp 121–171
[11] Wolk F, Yamazaki H, Seuront L and Lueck R G 2002 *J. Atm. Ocean. Tech.* **19** 780–793
[12] Jenkins W, Lott D, Cahill K, Curtice J and Landry P 2010 IOCCP Report No. 14: Sampling and measuring helium isotopes and tritium in seawater
[13] Stanley R H R, Baschek B, Lott D E and Jenkins W J 2009 *Geochem. Geophys. Geosyst.* **10** Q05008
[14] Hamme R and Emerson S 2004 *Deep-Sea Res. I* **51** 1517–1528
[15] Weiss R F 1971 *J. Chem. Eng. Data* **16** 235–241
[16] Weiss R F and Kyser T K 1978 *J. Chem. Eng. Data* **23** 69–72
[17] Wood D and Caputi R 1966 *Technical report. U.S. Naval Radiological Defense Laboratory, San Francisco, CA*
[18] Aoki N and Makide Y 2005 *Chemistry Letters* **34** 1396–1397
[19] Jähne B, Münnich K O, Bösinger R, Dutzi A, Huber W and Libner P 1987 *J. Geophys. Res.* **92** 1937–1949
[20] Manning C C and Nicholson D P 2016 gas_toolbox: Updated release for Manning et al. GTWS-7 proceedings. Zenodo. `http://dx.doi.org/10.5281/zenodo.46569`
[21] Hsu S, Meindl E A and Gilhousen D B 1994 *J. Appl. Meteor.* **33** 757–765
[22] Dickson A G, Sabine C L, Christian J R *et al.* 2007 *Guide to best practices for ocean CO_2 measurements* PICES Special Publication 3 (North Pacific Marine Science Organization) chap 5
[23] Fitzwater S E, Johnson K S, Elrod V A, Ryan J P, Coletti L J, Tanner S J, Gordon R M and Chavez F P 2003 *Continental Shelf Research* **23** 1523–1544
[24] Stanley R H and Jenkins W J 2013 *The Noble Gases as Geochemical Tracers* (Springer) pp 55–79
[25] Liang J H, Deutsch C, McWilliams J C, Baschek B, Sullivan P P and Chiba D 2013 *Global Biogeochem. Cy.* **27** 894–905
[26] Sweeney C, Gloor E, Jacobson A R, Key R M, McKinley G, Sarmiento J L and Wanninkhof R 2007 *Global Biogeochem. Cy.* **21**
[27] Wanninkhof R 2014 *Limnol. Oceanogr. Meth.* **12** 351–362
[28] Ho D T, Law C S, Smith M J, Schlosser P, Harvey M and Hill P 2006 *Geophys. Res. Lett.* **33** 16611
[29] Naegler T, Ciais P, Rodgers K and Levin I 2006 *Geophys. Res. Lett.* **33**
[30] Wanninkhof R and McGillis W R 1999 *Geophys. Res. Lett.* **26** 1889–1892
[31] Sullivan P P, Romero L, McWilliams J C and Melville W K 2012 *J. Phys. Oceanogr.* **42** 1959–1980
[32] Large W and Pond S 1981 *J. Phys. Oceanogr.* **11** 324–336
[33] McDougall T and Barker P 2011 Getting started with TEOS-10 and the Gibbs Seawater (GSW) Oceanographic Toolbox. SCOR/IAPSO WG127

Gas exchange of CO_2 and O_2 in partially ice-covered regions of the Arctic Ocean investigated using *in situ* sensors

F Islam[1], M DeGrandpre[1], C Beatty[1], R Krishfield[2] and J Toole[2]

[1]Department of Chemistry and Biochemistry, University of Montana, Missoula, MT
[2]Woods Hole Oceanographic Institution, Woods Hole, MA

Email: michael.degrandpre@umontana.edu

Sea surface CO_2 dynamics are not well characterized in the Arctic Ocean (AO). Most data are from ship-based studies during the low-ice period (May–September) and obtained from near shore areas because of accessibility. More CO_2 data are needed to improve models for predicting the future of the carbon cycle in the region and its relationship to ocean acidification. Air-sea gas exchange rates are complicated by the presence of ice. Consequently gas exchange rates have a larger uncertainty in the AO compared to other ocean regions. To provide more information about CO_2 dynamics and gas exchange in the AO, *in situ* time-series data have been collected from the Canada Basin during late summer to autumn of 2012. Partial pressure of CO_2 (pCO_2), dissolved O_2 (DO) concentration, temperature, and salinity were measured at ~6-m depth under little ice and multi-year ice on two ice-tethered profilers (ITPs) for 40–50 days. The pCO_2 levels were always below atmospheric saturation whereas DO was almost always slightly above saturation. Although the two ITPs were on an average only 222 km apart, one was further south and had 14 ± 12% ice cover; whereas the more northern ITP had 63 ± 16% ice cover. Consequently the two data sets differed significantly in external forcings. Modeled variability of CO_2 and DO estimate that gas exchange would significantly alter sea surface pCO_2 in the low ice cover record but minimally in the more extensively ice-covered region. If these conditions extended over the entire AO, the total uptake of atmospheric CO_2 would be 28.6 Tg C yr^{-1} and 15.4 Tg C yr^{-1} under low and high ice-covered conditions, respectively.

1. Introduction

The rapid loss of ice covering the Arctic Ocean (AO) is increasing exchange of CO_2 and O_2 between the atmosphere and AO sea surface. There is evidence that loss of ice has already increased the partial pressure of CO_2 (pCO_2) levels in AO surface waters (Cai *et al.* 2010; Else *et al.* 2013). While the additional uptake of atmospheric CO_2 may lessen global warming, ocean acidification might be accelerated as ice cover decreases, potentially resulting in a rapid pH decrease (Steinacher *et al.* 2009). It is uncertain, however, if the AO will absorb significant CO_2 under seasonal ice-free conditions. Warming of the shallow isolated surface water and weak biological CO_2 drawdown are hypothesized to limit CO_2 invasion, keeping the ice-free central AO basins from becoming large atmospheric CO_2 sinks (Cai *et al.* 2010).

Content from this work may be used under the terms of the Creative Commons Attribution 3.0 licence. Any further distribution of this work must maintain attribution to the author(s) and the title of the work, journal citation and DOI.
Published under licence by IOP Publishing Ltd

The lack of pCO_2 and other supporting data has limited our ability to predict how the carbon cycle will respond to diminished ice cover. Sea surface pCO_2 data sets from the central AO are scarce with most Arctic studies focused on the AO continental shelves (Anderson *et al.* 2009; Bates *et al.* 2006; Murata and Takizawa 2003; Kaltin and Anderson 2005). Those studies that have focused on the central ocean basins have generally found that pCO_2 is undersaturated, ranging from 240–350 μatm (Anderson and Kaltin 2001; Bates 2006; Bates *et al.* 2006; Cai *et al.* 2010; Else *et al.* 2013; Robbins *et al.* 2013). Almost all of these studies were during the low-ice periods in the summer. There continue to be significant gaps in our understanding of CO_2 dynamics in the AO, especially under ice. To examine pCO_2 and dissolved O_2 (DO) variability in the AO, in late August 2012 we deployed pCO_2, DO, temperature, salinity and bio-optical sensors on ice-tethered profilers (ITPs) in the Canada Basin. In 2012, the AO reached the lowest seasonal ice extent since satellites began quantifying ice coverage in 1979 (Parkinson *et al.* 2013; Zhang *et al.* 2013). The loss of ice is greater in the Canada Basin compared to the other basins in the AO (McLaughlin *et al.* 2011) and therefore the carbon cycle changes are expected to be the greatest in this region. The two ITPs were deployed in densely and sparsely ice-covered locations, providing a unique opportunity to compare CO_2 and DO variability under these different conditions. In this paper we evaluate the gas exchange contribution to these two time series and estimate total air-sea gas flux for the AO.

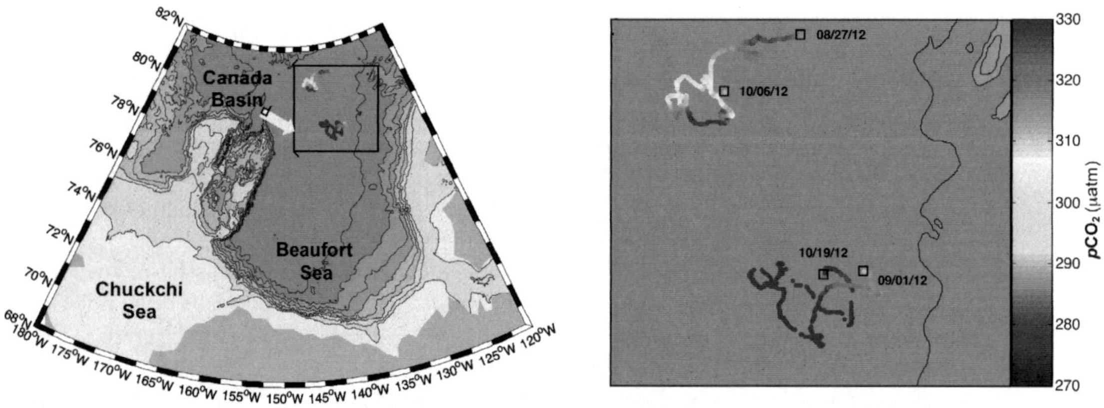

Figure 1. Arctic marine regions (left) near the study area (inset) are shown with the ITP drift track overlaid. The study area is magnified on the right. Rectangular boxes labeled with dates highlight measurement start and end points. The pCO_2 values at different points over the drift track are shown in color. The most northerly track corresponds to ITP-65 and the southern track to ITP-64. The sensors were deployed ~245 km away from each other. Plotting software is courtesy of Bill Williams (Institute of Ocean Sciences, Canada).

2. Methods

The Canada Basin is the largest sub-basin of the Arctic Ocean, extending approximately 1100 km from the Beaufort Sea shelf to the Canadian Archipelago (figure 1). Our instrumentation was deployed at ~6-m depth on two ITPs (ITP-64 and ITP-65) (Toole *et al.* 2011; Krishfield *et al.* 2008) in the Canada Basin from the Canadian Coast Guard Icebreaker *Louis S. St. Laurent*. ITP-65 was deployed on a 1.5-m thick ice floe at 80° 53.4′ N, 137° 25.8′ W and ITP-64 was deployed in open water at 78° 46.5′ N, 136° 39.8′ W. ITP-64 and ITP-65 traveled 513 and 575 km, respectively, during the period when the sensors transmitted data. Our sensors were deployed ~ two weeks before the ice extent reached its minimum on September 16.

The ITPs are described in Krishfield *et al.* (2008). Autonomous CO_2 sensors (SAMI-CO_2, Sunburst Sensors, LLC) (DeGrandpre *et al.* 1995) were connected to the ITP wire at ~6-m depth and recorded pCO_2 at 2-hour intervals. The instruments also included DO sensors (Optode 4175, Aanderaa

Data Instruments) and CTDs (SBE37SI, Sea-Bird Electronics Inc.). The SAMIs transmitted data to the surface controller through an inductive modem interface (SBE UIMM, Sea-Bird Electronics Inc.). SAMI and optode data were quality controlled using shipboard measurements and data collected at the time of deployment. For unknown reasons, the SAMI inductive modem communications failed after 40–50 days. The transmitted data are available at the Woods Hole Oceanographic Institution (WHOI) website (http://www.whoi.edu/page.do?pid=23099) and the Arctic data repository ACADIS.

A variety of other data were obtained from different sources. The wind speed data near ITP locations were downloaded from the European Center for Medium range Weather Forecasting (ECMWF) website: http://apps.ecmwf.int/datasets/data/interim-full-daily. Barometric pressure and air temperature data were obtained from the Ice-Mass Balance (IMB) Buoy website: http://imb.erdc.dren.mil/2012L.htm. Atmospheric CO_2 measured at Barrow, Alaska was obtained from the NOAA Earth System Research Laboratory (ESRL) website: http://www.esrl.noaa.gov/gmd/dv/data. Ice-coverage data were derived from Nimbus-7 SMMR and DMSP SSM/I-SSMIS Passive Microwave Data compiled by the National Snow and Ice Data Center (NSIDC) (Cavalieri *et al.* 1996) (http://nsidc.org/data).

Alkalinity was derived from a salinity–total alkalinity (A_{Tsalin}) relationship in the AO (Yamamoto-Kawai *et al.* 2005). Total dissolved inorganic carbon (DIC) was calculated using the measured pCO_2 and A_{Tsalin} using the program CO2SYS (van Heuven *et al.* 2011). The mixed layer depth (MLD) was calculated from depth-resolved density and temperature, with MLD as the depth where the density difference from the SAMI depth is +0.3 kg m^{-3}. We used this density difference because it gives computed MLDs that best match *in situ* MLDs when observed visually based on the water column density profile. This value is close to the value of 0.26 kg m^{-3} reported by Timmermans *et al.* (2010) using the same approach for the central Canada Basin. Water was well mixed above 6 m based on ship CTD casts obtained during deployment (not shown).

The flux of DIC and DO originates primarily from four different processes in the AO, as shown in equation 1.

$$MLD \times dC/dt = F_{gasex} + F_{NCP} + F_{mix} + F_{brine} \quad (1)$$

where dC is the difference in DIC or DO values between two measurements over time (dt), and F_{gasex}, F_{NCP}, F_{mix}, and F_{brine} are the fluxes due to gas transfer across the air-water interface, net community production (NCP), vertical mixing and advection, and brine rejection during ice formation, respectively. We assume $CaCO_3$ dissolution and formation are not significant. In this evaluation, we focus on calculation of F_{gasex}.

The F_{gasex} of CO_2 and DO across the air-water interface was calculated using the diffusive boundary layer model:

$$F_{gasex} = k\,\Delta C\,f \quad (2)$$

where k is the gas transfer velocity, ΔC is the concentration gradient across the air-sea interface, and f is the fraction of open water. k was estimated using the wind speed relationship in Wanninkhof (2014) and adjusted for different temperatures and gases (i.e., CO_2, DO). The bubble flux equation in Stanley *et al.* (2009) was included for the DO fluxes to account for O_2 supersaturation due to complete bubble injection. We have not used relationships that account for other processes such as ice-sea surface turbulence in partially ice-covered oceans (Loose *et al.* 2014) but rather explore the range of air-sea CO_2 and DO fluxes by comparing models with the recorded sea ice extent and without ice, as discussed below.

The mass balance (equation 1) was used in a spreadsheet model for carbonate system calculations (DeGrandpre *et al.* 2004). The model takes the initial DIC, a constant A_T and the *in situ* temperature and salinity in the surface seawater as the initial input. The DIC was incremented at each time step for contributions from air-sea gas exchange using the calculated pCO_2. The new pCO_2 was then calculated from the constant A_T, *in situ* temperature and salinity, and the new DIC. This calculation gives the modeled variability of pCO_2 from gas exchange and temperature. A similar approach was used in the DO model. Both pCO_2 and DO gas exchange models were calculated assuming ice-free conditions as well, with the goal to show to what extent AO ice-coverage controls pCO_2 and DO variability.

3. Results

The 49-day time-series data for ITP-64 and 41-day time-series data for ITP-65 are shown in figure 2. Gaps in the data correspond to instrument data communication dropouts between the SAMI and inductive modem. The average ice cover at ITP-64 and ITP-65 was 14 ± 12% and 63.2 ± 15.8%, respectively (figure 2). The average pCO_2 at ITP-64 and ITP-65 was 334 ± 9 μatm and 304 ± 16 μatm, respectively. DO showed very little variability at both locations with a mean of 392 ± 2 μmol kg^{-1} at ITP-64 and 385 ± 2 μmol kg^{-1} at ITP-65 (figure 2). Note that the diurnal cycles are distinct early in ITP-64 DO data but are not apparent in the ITP-65 DO data. Diurnal cycles in DIC at ITP-64 are also evident once the long-term trend is removed with a bandpass filter (Islam *et al.* 2016).

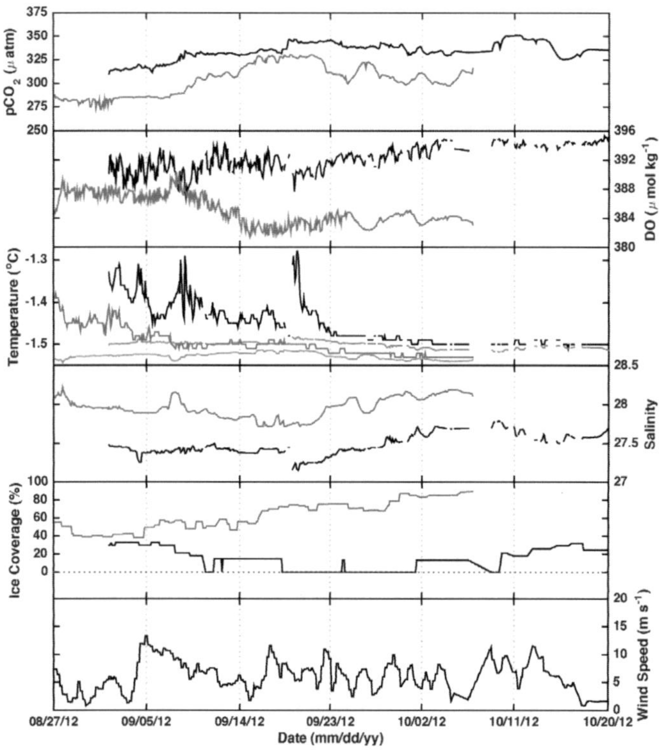

Figure 2. Biogeochemical and physical time-series data collected at ~6 m on ITP-64 (black) and ITP-65 (red). ITP-64 was deployed in open water whereas ITP-65 was deployed under multi-year ice (ice coverage is shown in the fifth panel). Gaps in the data correspond to instrument communication dropouts. The light black line and light red line in the temperature plot refer to freezing temperature of water at ITP-64 and ITP-65, respectively (third panel). Wind speed is shown in the bottom panel. The same wind speed was used for both locations (see section 2 for more information about data sources).

4. Discussion

4.1. ITP-64

The modeled pCO_2 trend due to gas exchange initially closely follows the measured pCO_2, but undershoots after 9/9/12 (figure 3). The modeled gas exchange with no ice is very similar because there is very little ice coverage at this location (figure 2). Gas exchange without bubble injection drives modeled DO quickly to saturation and underpredicts the observed DO levels (figure 3). Bubble fluxes account for part of the supersaturation (red curve). The pCO_2 and DO model deviations from the *in situ* data indicate that there are other processes altering the pCO_2 and DO levels. These other sources of variability are evaluated in Islam *et al.* (2016).

4.2. ITP-65

The contribution from gas exchange to the ITP-65 pCO_2 variability is relatively small (figure 4) and would be significantly larger if no ice was present (dashed blue curve). The modeled gas exchange departs significantly from the observations. Much of this variability is likely due to spatial pCO_2 gradients as mentioned above. In contrast, gas exchange explains almost all of the DO variability (figure 4). Bubble fluxes add a slight supersaturation. The underestimate of pCO_2 and overestimate of DO can be attributed, in part, to biological processes, in this case, net respiration (Islam *et al.* 2016).

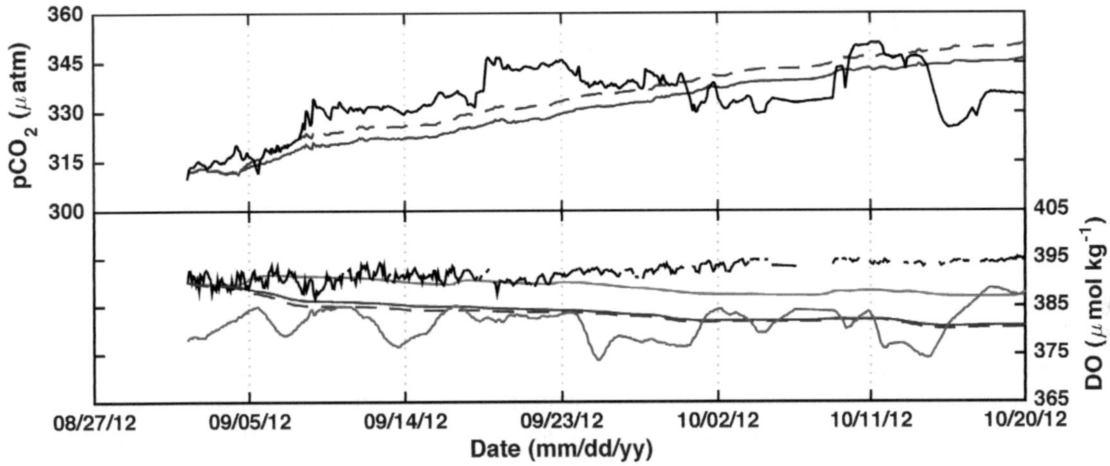

Figure 3. ITP-64 pCO_2 and DO data (black) and modeled gas exchange using the ice cover (blue solid) and without ice cover (blue dashed). Modeled DO gas exchange including complete bubble injection is shown in red. The green line corresponds to the equilibrium concentration of O_2 at *in situ* temperature, salinity and atmospheric pressure. The pCO_2 data were calculated using *in situ* temperature and salinity, constant A_T, and with DIC incremented due to gas exchange (see text for more details).

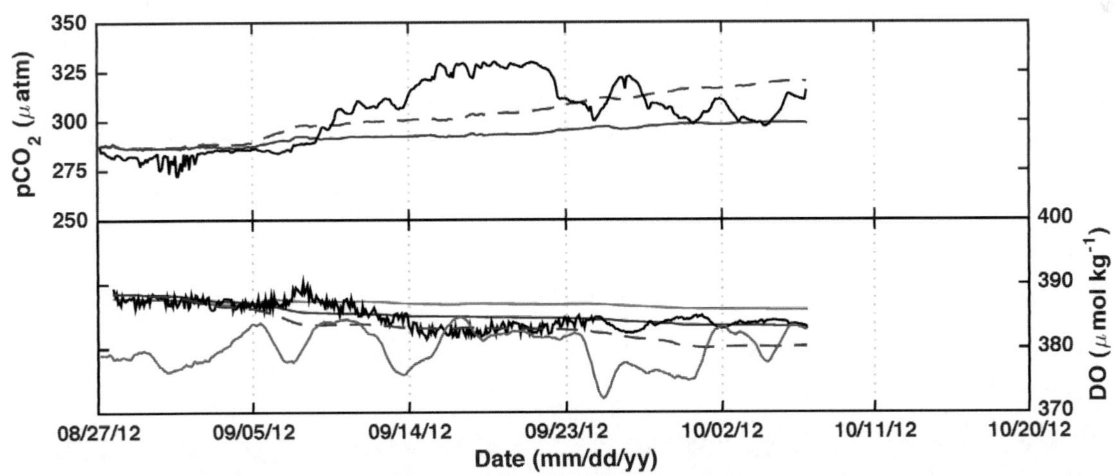

Figure 4. ITP-65 pCO_2 and DO data (black) and modeled gas exchange using the ice cover (blue solid) shown in figure 2 and without ice cover (blue dashed). Modeled DO gas exchange including complete bubble injection is shown in red. The green line corresponds to the equilibrium concentration of O_2 at *in situ* temperature, salinity and atmospheric pressure. The pCO_2 data were calculated using *in situ* temperature and salinity, a constant A_T, and with DIC incremented due to gas exchange (see text for more details). The x-axis scale is selected to match the other time-series figures.

4.3. Air-sea CO_2 fluxes

The CO_2 flux estimates from our data fall within the range found by others for the Canada Basin and nearby regions. The CO_2 flux at ITP-64 was –4.8 ± 3.8 mmol m^{-2} d^{-1} and –2.5 ± 2.6 mmol m^{-2} d^{-1} for ITP-65 (negative values indicate uptake of CO_2). Murata and Takizawa (2003) report that the CO_2 flux during the summers of 1998–2000 was –12 mmol m^{-2} d^{-1} in the western Beaufort Sea. Bates *et al.* (2006) estimated that during summertime the fluxes in the Canada Basin averaged –55 mmol m^{-2} d^{-1}. Assuming $p$$CO_2$ and ice coverage characteristic of the means found at the ITPs extended over a 90-day period and extrapolating over the entire Canada Basin (4,489,000 km^2) (Macdonald *et al.* 2009), the CO_2 flux is estimated to be –12.5 Tg C yr^{-1} under ITP-64 conditions (14 ± 12% ice cover) and –6.5 Tg C yr^{-1} under ITP-65 conditions (63 ± 16% ice cover). Bates *et al.* (2006) estimate the annual CO_2 flux in the Canada Basin to be in the range of –6 to –19 Tg C yr^{-1}. Our annual fluxes at ITP-64 are also similar to those estimated by Evans *et al.* (2015) for the western Arctic coastal ocean. When the entire AO area of 10,700,000 km^2 (Macdonald *et al.* 2009) is considered, our flux estimates are –28.6 Tg C yr^{-1} and –15.4 Tg C yr^{-1} under ITP-64 and ITP-65 conditions, respectively. The ITP-65 values are on the high end of the flux estimates reported by Bates *et al.* (2011) for the central Arctic basins likely because of the low ice cover during 2012.

5. Conclusions

This study documented the $p$$CO_2$ levels in marginally and partially ice-covered regions in the Canada Basin. The $p$$CO_2$ and DO were significantly undersaturated and slightly supersaturated, respectively. In low ice cover, gas exchange had a significant influence on $p$$CO_2$ but was a minor contributor in the more highly ice-covered region. If ice cover continues to diminish in the AO, we expect the CO_2 accumulation will outpace the sinks of CO_2 in the AO sea surface because of the low NCP (export production), isolation of the surface layer because of strong stratification and the long residence time of the surface water. Therefore, $p$$CO_2$ levels may continue to rise towards saturation from year to year and the net air-sea flux observed in this study will diminish. Continued monitoring of $p$$CO_2$ in the AO is required to document these predicted trends.

Acknowledgments

We thank Kris Newhall, Steve Lambert, Jim Dunne, and Jeff O'Brien from Woods Hole Oceanographic Institution, Jim Beck from Sunburst Sensors, and Brittany Peterson from the University of Montana. Rachel Stanley provided helpful discussions regarding bubble injection. We also thank the captain, crew and chief scientist (Bill Williams) of the CCGS *Louis S. St. Laurent*. Funding was provided by the U.S. National Science Foundation Arctic Observing Network (ARC – 1107346).

References

Anderson L G *et al* 2009 *Geophys. Res. Lett.* **36** L20601
Anderson L G and Kaltin S 2001 *Polar Res.* **20** 225
Bates N R *et al* 2011*Oceanography* **24** 186
Bates N R 2006 *J. Geophys. Res.* **111** C10013
Bates N R *et al* 2006 *Geophys. Res. Lett.* **33** L23609
Bates N R *et al* 2005 *Deep-Sea Res. II* **52** 3303
Cai W *et al* 2010 *Science* **329** 556
Cavalieri D *et al* 1996*Sea Ice Concentrations from Nimbus-7 SMMR and DMSP SSM/I-SSMIS Passive Microwave Data, Version 1* (Boulder, Colorado USA: NASA DAAC at the National Snow and Ice Data Center)
DeGrandpre M D *et al* 2004 *J. Geophys. Res.* **109** C08S07
DeGrandpre M D *et al* 1995 *Limnol. Oceanog.* **40** 969
Else G T *et al* 2013 *Geophys. Res. Lett.* **40** 1132
Evans W *et al* 2015 *Global Biogeochem. Cycles* **29** 1

IOP Conf. Series: Earth and Environmental Science **35** (2016) 012018 doi:10.1088/1755-1315/35/1/012018

Islam F *et al* 2016 *J. Geophys. Res.* under review
Loose B *et al* 2014*Ocean Sci.* **10** 17
Kaltin S and Anderson L G 2005 *Mar. Chem.* **94** 67
Krishfield R *et al* 2008 *J. Atmos. Ocean. Technol.* **25** 2091
Macdonald R W *et al* 2009 *Carbon and Nutrient Fluxes in Continental Margins: A Global Synthesis* eds Liu K K *et al* (New York: Springer) 291
McLaughlin F *et al* 2011 *Oceanography* **24** 146
Murata A and Takizawa T 2003 *Cont. Shelf Res.* **23** 753
Parkinson C L and Comiso J C 2013 *Geophys. Res. Lett.* **40** 1356
Robbins L L *et al* 2013 *PLoS ONE* **8** e73796
Stanley R H R *et al* 2009 *J. Geophys. Res.* **114** doi:10.1029/2009JC005396
Steinacher M *et al* 2009 *Biogeosciences* **6** 515
Timmermans M-L *et al* 2010 *J. Atmos. Ocean. Technol.* **27** 1936
Toole J M *et al* 2011*Oceanography* **24** 126
van Heuven S *et al* 2011 *MATLAB Program Developed for CO_2 System Calculations* ORNL/CDIAC-105b (Oak Ridge, Tennessee: Oak Ridge National Laboratory)
Wanninkhof R 2014 *Limnol. Oceanogr. Methods* **12** 351
Woolf D K and Thorpe S A 1991 *J. Mar. Res.* **49** 435
Yamamoto-Kawai M *et al* 2005 *J. Geophys. Res.* **110** C10003
Zhang J *et al* 2013 *Geophys. Res. Lett.* **40** 1-7

Noble gas tracers of ventilation during deep-water formation in the Weddell Sea

D P Nicholson[1], S Khatiwala[2] and P Heimbach[3,4]

[1]Marine Chemistry and Geochemistry Department, Woods Hole Oceanographic Institution, 266 Woods Hole Road, Woods Hole, MA 02543, USA
[2]Department of Earth Sciences, University of Oxford, South Parks Road, Oxford OX1 3AN, UK
[3]Institute for Computational Engineering and Sciences & Jackson School of Geosciences, The University of Texas at Austin, 201 East 24th Street, POB 4.232, Austin, TX 78712, USA
[4]Department of Earth, Atmospheric and Planetary Sciences, Massachusetts Institute of Technology, 77 Massachusetts Avenue, Cambridge, MA, 02139, USA

E-mail: dnicholson@whoi.edu

Abstract. To explore the dynamics and implications of incomplete air-sea equilibration during the formation of abyssal water masses, we simulated noble gases in the Estimating the Circulation & Climate of the Ocean (ECCO) global ocean state estimate. A novel computation approach utilizing a matrix-free Newton–Krylov (MFNK) scheme was applied to quickly compute the periodic seasonal solutions for noble gas tracers. MFNK allows for quick computation of a cyclo-stationary solution for tracers (i.e., a spun-up, repeating seasonal cycle), which would otherwise be computationally infeasible due to the long time scale of dynamic adjustment of the abyssal ocean (1000's of years). A suite of experiments isolates individual processes, including atmospheric pressure effects, the solubility pump and air-sea bubble fluxes. In addition to these modeled processes, a volumetric contribution of 0.28 ± 0.07% of glacial melt water is required to reconcile deep-water observations in the Weddell Sea. Another primary finding of our work is that the saturation anomaly of heavy noble gases in model simulations is in excess of two-fold more negative than is suggested from Weddell Sea observations. This result suggests that model water masses are insufficiently ventilated prior to subduction and thus there is insufficient communication between atmosphere and ocean at high latitudes. The discrepancy between noble gas observations and ECCO simulations highlights that important inadequacies remain in how we model high-latitude ventilation with large implications for the oceanic uptake and storage of carbon.

1. Introduction

The vast abyssal ocean is ventilated by a few limited surface regions of deep-water formation including the Labrador Sea and Weddell Sea [1]. Thus, ocean storage of dissolved inorganic carbon (DIC) depends critically on air-sea exchange processes in these regions [2–4]. The zonally and vertically integrated poleward volume transport results in the meridional overturning circulation, which is composed of an upper (northern) and lower (southern) cell resulting in upwelling in the

[1] To whom any correspondence should be addressed.

Content from this work may be used under the terms of the Creative Commons Attribution 3.0 licence. Any further distribution of this work must maintain attribution to the author(s) and the title of the work, journal citation and DOI.
Published under licence by IOP Publishing Ltd

Southern Ocean, a portion of which returns to depth, contributing to Antarctic Bottom Water formation (figure 1). This portion of the flow experiences incomplete ventilation while at the (often sea-ice covered) surface in the Southern Ocean [5]. The degree to which Southern Ocean surface waters equilibrate before subducting to form either AABW or intermediate water masses has a large impact on the efficiency with which preformed nutrients are utilized [6] and the ocean-atmosphere partitioning of CO_2. The resultant degree of partial equilibration dramatically influences the DIC capacity of the ocean, atmospheric pCO_2 and global climate [3,7].

Noble gases record this incomplete equilibration as evidenced by observed undersaturation of heavy noble gases (Xe, Kr) in the deep ocean [8–10]. While the inorganic carbon system is subject to complex interactions of biological and physical mechanisms, the noble gases are abiotic and thus useful for isolating and understanding the physical dynamics of air-sea gas exchange and ventilation. Here, we combine numerical simulations with noble gas observations to examine signatures of incomplete equilibration in the deep Weddell Sea.

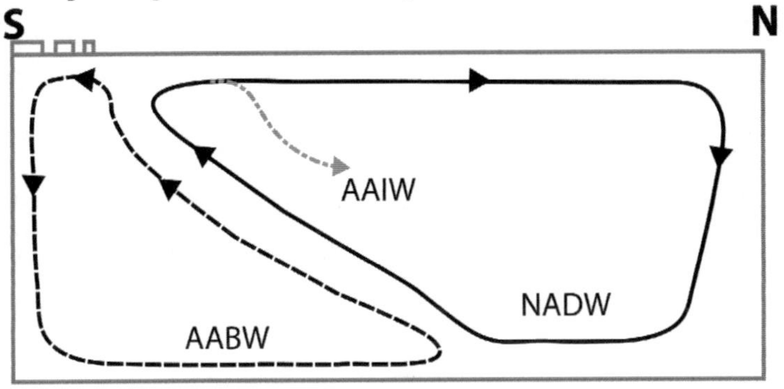

Figure 1. Schematic of the meridional overturning circulation, showing North Atlantic Deep Water (NADW), Antarctic Bottom Water (AABW) and Antarctic Intermediate Water (AAIW). Air-sea equilibration time in the Southern Ocean is limited by residence time, partial sea-ice cover and deep winter mixed layer depths.

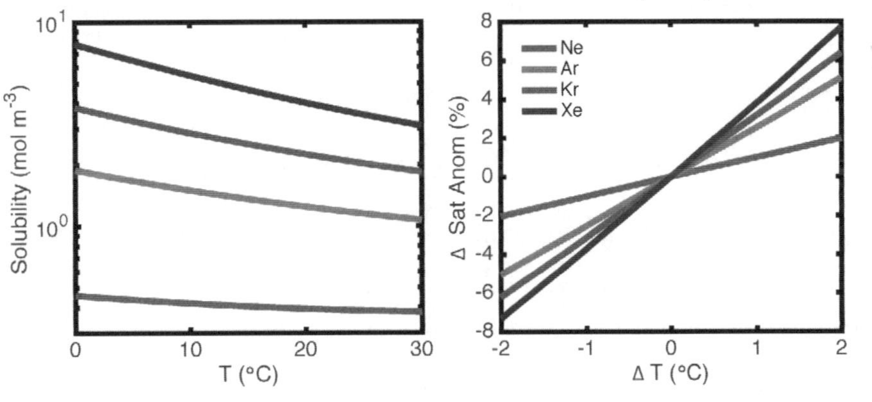

Figure 2. Solubility of noble gases (left) and temperature dependence of solubility at 5°C (right). The right panel shows the resulting change in saturation state given a change in temperature without change in gas concentration.

2. Background

The noble gases are useful tracers of air-sea processes because they have a range of physical properties (figure 2). Each has a well-known boundary condition (a well-mixed tropospheric value) and with the exception of He, they have no sources or sinks in the ocean interior [11]. These tracers have previously been applied to understand a range of oceanic processes, including air-sea bubble fluxes [12,13], diapycnal mixing [14,15], ventilation during deepwater formation [8, 9] and glacial melting [10]. Generally, the concentration of noble gases is within a few percent of solubility equilibrium, with small deviations induced by a range of physical processes of interest. Fortunately, such saturation anomalies (ΔC) can analytically be determined precisely and are denoted as:

$$\Delta C = \left(\frac{[C]-[C]_{eq}}{[C]_{eq}} \right) \tag{1}$$

where $[C]$ is the measured concentration of gas C, and $[C]_{eq}$ is the temperature and salinity-dependent equilibrium solubility at 1 atm and saturated water vapour pressure [16–18].

The air-sea interface is the dominant source and sink for dissolved noble gases in the ocean. The air-sea flux of sparingly soluble gases such as the noble gases is distinctly influenced by bubble-mediated processes [19–21] and thus an explicit bubble parameterization is warranted. Following previous noble gas studies [12, 13, 22], we parameterize air-sea flux to include a term for the surface diffusive component, as well as two bubble-mediated flux terms such that the net air to sea flux (F_{as}) is:

$$F_{as} = F_{surf} + F_{bub} \tag{2}$$

where:

$$F_{surf} = -k_c(1 - f_{ice})(\delta C^P) \tag{3}$$

where k_C is the diffusive gas transfer velocity (m s^{-1}) and δC^P (mol m^{-3}) is the concentration anomaly corrected for local atmospheric pressure:

$$\delta C^P = [C] - [C]_{eq} \left(\frac{p_{slp} - p_w}{p_{ref} - p_{wsat}} \right) \tag{4}$$

where p_{slp} (atm) is sea level pressure, p_w (atm) is water vapour pressure at the sea surface, p_{ref} (atm) is standard atmospheric pressure, and p_{wsat} (atm) is water vapour pressure at 100% relative humidity,

$$F_{bub} = (1 - f_{ice}) U \chi_C \left(\frac{p_{slp}}{p_{ref}} \right) \left(A_{inj} + A_{ex} \left(\frac{D_C}{D_0} \right)^{0.5} \left(\frac{\alpha_C}{\alpha_0} \right) \right) \tag{5}$$

where $U = ((u_{10} - 2.27)/u_0)^3$ for $u_{10} > 2.27$ and otherwise $U = 0$. A_{inj} and A_{ex} are empirical constants, χ_C is the atmospheric mixing ratio of C, u_{10} is 10-m wind velocity (m s^{-1}), D_C is gas diffusivity (m^2 s^{-1}) and α_C is gas solubility (mol m^{-3} atm^{-1}). D_0, α_0 and u_0 are unit-value normalization constants with values of 1 m^2 s^{-1}, 1 mol m^{-3} atm^{-1} and 1 m s^{-1}, respectively.

2.1. Quasi-steady state

Given a steady wind speed and rate of cooling a surface water parcel will reach a quasi-steady state [14] such that the saturation anomaly is stable (i.e., the above equations are solved for the case where $d(\Delta C^P)/dt = 0$ and $\Delta C^P = \delta C^P/[C]_{eq}$):

$$\Delta C^P = \frac{-H}{c_p k_c(1-f_{ice})} \frac{1}{[C]_{eq}} \frac{d[C]_{eq}}{dT} + \frac{U}{k_c} \left(\frac{p_{slp}}{p_{ref}} \right) \left(\left(\frac{A_{inj}}{\alpha_C} \right) + A_{ex} \left(\frac{D_C}{D_0} \right)^{0.5} \right) \tag{6}$$

where H (W m^{-2}) is air-sea heat flux (positive is heat lost to the atmosphere), c_p (J m^{-3} K^{-1}) is the heat capacity and T (°C) is mixed layer temperature. The above quasi-steady-state equation is composed of a cooling/warming driven anomaly and a bubble-mediated anomaly. The magnitude of the temperature induced surface anomaly is larger for the heavy noble gases (due to larger $d([C]_{eq})/dT$) while the bubble component is smaller for the heavy gases (due to larger α_C). Another notable implication of the above equation is that ice cover amplifies the saturation anomaly due to heating/cooling, but cancels out for bubble-induced anomaly.

The quasi-steady-state framework is useful to illustrate how saturation anomaly is composed of a surface component driven by air-sea heat flux and a bubble-mediated component. Atmospheric pressure, which typically can reach 2–3% below 1 atm at high latitude also contributes to ΔC.

Table 1. Description and units for equation terms

Abbreviation	Description	Units
$[C]$	Dissolved concentration of gas	mol m^{-3}
$[C]_{eq}$	Equilibrium $[C]$ *w.r.t.* 1 atm and saturating water vapour pressure	mol m^{-3}
ΔC	Gas saturation anomaly	unitless
F_{as}	Total Air-sea gas flux	mol m^{-2} s^{-1}
F_{surf}	Surface air-sea gas flux	mol m^{-2} s^{-1}
F_{bub}	Bubble mediated air-sea gas flux	mol m^{-2} s^{-1}
k_c	Gas transfer piston velocity	m s^{-1}
f_{ice}	Fraction ice cover	unitless
δC^P	Saturation anomaly relative to local atmospheric pressure	mol m^{-3}
p_{slp}	Sea level pressure	atm
p_w	Water vapour pressure	atm
p_{wsat}	Saturation water vapour pressure	atm
χ_C	Dry atmospheric molar mixing ratio	mol/mol
A_{inj}	Emperical constant for bubble injection	mol m^{-2} s^{-1}
A_{ex}	Emperical constant for bubble exchange	mol m^{-2} s^{-1}
u_{10}	10-meter wind speed	m s^{-1}
D_C	Gas diffusivity	m^2 s^{-1}
α_C	Gas solubility	mol m^{-3} atm^{-1}
H	Net air-sea heat flux (negative for ocean cooling)	W m^{-2}
c_p	Seawater heat capacity	J m^{-3} K^{-1}
T	seawater temperature	°C

3. Model methods

3.1. The noble gases in ECCO

The circulation used in this study is based on an ocean state estimate provided by the Estimating the Circulation & Climate of the Ocean (ECCO) consortium [23]. The ECCO ocean state estimate (ECCO version 2) was achieved by adjusting the air-sea fluxes of heat, momentum and freshwater in the MIT ocean general circulation model [24] through data assimilation [25, 26]. This procedure results in a dynamically consistent estimate of ocean circulation and hydrography over the assimilation period. The model has a horizontal resolution of 1° with 23 vertical levels. The actual tracer simulations were performed using the transport matrix method (TMM), an offline scheme for simulation of ocean biogeochemical tracers [27, 28]. Furthermore, in order to efficiently spin-up each of the large number of runs to equilibrium, a matrix-free Newton–Krylov method was used [29]. This approach is significantly more efficient than simply integrating the model to equilibrium. Monthly mean transport matrices, representing a climatology over the 1992–2004 assimilation period, were extracted from the ECCO state estimate and used to carry out the tracer simulations. The ECCO transport matrices along with the TMM code are available from *https://github.com/samarkhatiwala/tmm*. Our model solutions are thus comparable to a fully spun-up ocean with a seasonal cycle representative of the average seasonal cycle for the model training period (1992–2004).

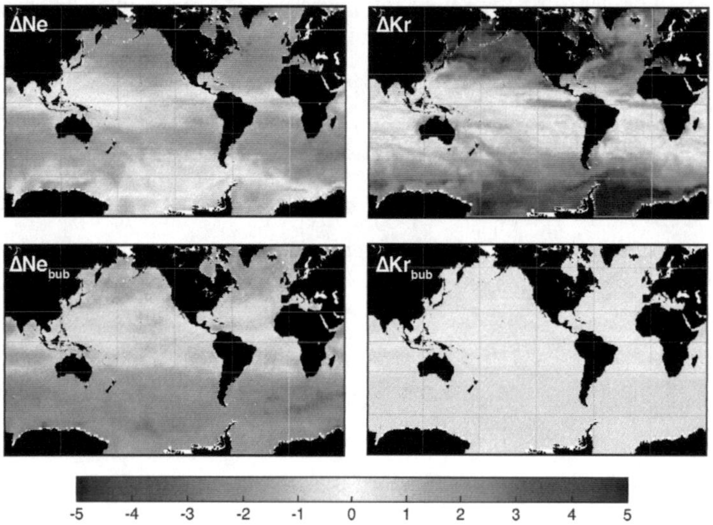

Figure 3. Modelled noble gas surface saturation anomalies shown for July for Ne (left) and Kr (right). Upper panels show total saturation anomaly while lower panels show the anomaly due to air-sea bubble flux. While bubble processes dominate surface ΔNe patterns they have a much smaller impact on ΔKr, which is dictated largely by saturation anomalies due to cooling and warming.

3.2. The noble gases in ECCO

Noble gases were simulated by applying equations (2)–(5) to dictate air-sea exchange. The gas transfer velocity was quadratic with wind speed using the parameterization of Sweeney *et al.* [30], and wind speed and sea level pressure taken from the 6-hourly CORE v2 dataset [31]. Gas diffusivities are calculated from Jahne *et al.* [32] and solubilities for Ne and Ar from Hamme and Emerson [16], for Kr from Weiss and Kyser [17] and for Xe from Wood and Caputi [18] adding a 2% correction as suggested by Hamme and Severinghaus [8].

Model simulations allow us to decompose noble gas saturation anomaly into a contribution from each relevant process. To do so, four simulations were performed for each noble gas consisting of (sim1) diffusive gas exchange only (A_{ex} and A_{inj} = 0), (sim2) diffusive gas exchange plus bubble injection (A_{ex} = 0), (sim3) diffusive gas exchange plus bubble exchange (A_{inj} = 0), and (sim4) all processes 'on'. Saturation anomaly due to a single process is calculated from the difference of two simulations, for example, $\Delta Ne_{inj} \approx \Delta Ne_{sim4} - \Delta Ne_{sim2}$ (figure 3).

4. Results

4.1. Air-sea bubble fluxes and atmospheric pressure

Similar to our previous work [13], but using a seasonally resolving and higher spatial resolution model, we constrained the magnitude of bubble injection and bubble exchange terms, A_{inj} and A_{ex}. All available Ne, Ar, Kr, Xe and N_2 observations from below 500 m were used in the inverse calculation. Best-fit values were $A_{inj} = 1.06 \times 10^{-9}$ and $A_{ex} = 2.19 \times 10^{-6}$. These values are 37% and 71% lower than previously published values for A_{inj} and A_{ex}, respectively [13]. Improvements in the underlying model, as well as the addition of newly published data contribute to the difference, but the primary cause for a lower bubble flux estimate is the use of higher-frequency wind forcing. In prior methods, monthly average winds were used and Rayleigh wind speed distributions were assumed to correct for non-linear averaging effects. In the current study, necessary non-linear wind speed forcing quantities $(u_{10}-2.27)^3$ and u_{10}^2 were calculated using 6-hourly winds, obviating the need for assumptions about wind speed distributions.

Beyond refining estimates of air-sea bubble fluxes, we constrain bubble fluxes here in order to remove the bubble signal from heavy noble gas distributions and focus on cooling driven undersaturation of the heavy noble gases. We find that for Kr and Xe, air sea bubble fluxes are a relatively minor component of observed saturation anomaly, contributing less than 1% saturation anomaly (figure 4). In the deep Weddell Sea, the combination of the two bubble mechanisms increase saturation anomalies for Ne, Ar, Kr and Xe by 2.1%, 0.7%, 0.5% and 0.3%, respectively. Consistently

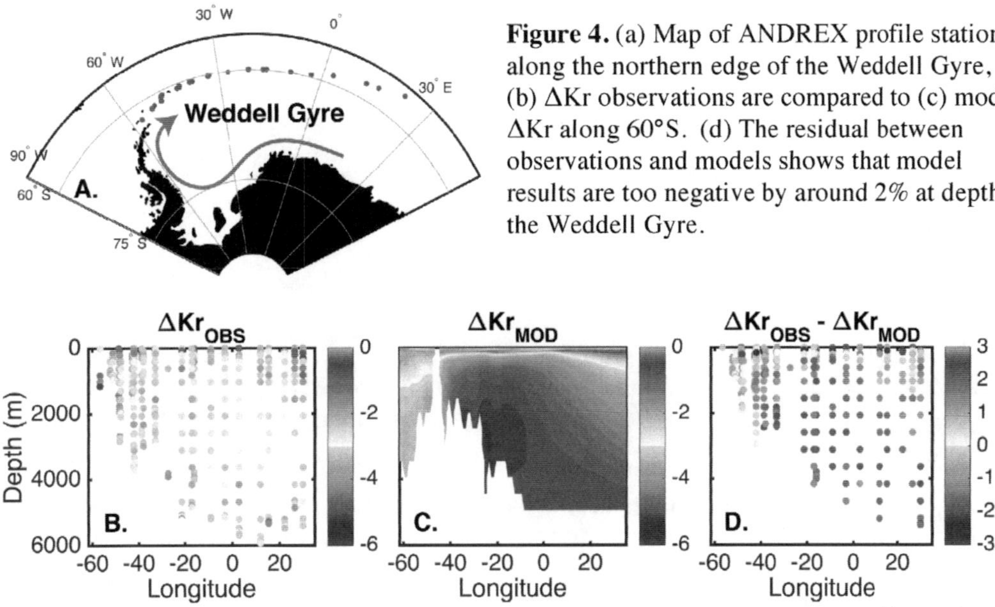

Figure 4. (a) Map of ANDREX profile stations along the northern edge of the Weddell Gyre, and (b) ΔKr observations are compared to (c) model ΔKr along 60°S. (d) The residual between observations and models shows that model results are too negative by around 2% at depth in the Weddell Gyre.

low atmospheric pressure around the Antarctic continent reduces the saturation anomaly for each gas by 1.6%.

4.2. Comparison to Weddell Sea observations

A recently published dataset [10] provides the first comprehensive observations that include the heavy noble gases in the Southern Ocean. The data were collected in April 2010 as part of the Antarctic Deep Water Rates of Export (ANDREX) project and consist of a quasi-zonal section through the Weddell gyre, approximately along the 60°S parallel from near the Drake Passage at 60°W to 30°E (figure 4). At the surface, there is significant seasonality in modelled gas saturation anomaly (supplementary figure S1) and observations are noisier, likely due to short term variability on time-scales shorter than the monthly-mean model results (figures S1 and 4). Compared to ANDREX observations the deep Weddell gyre is significantly more undersaturated in the model. For a light noble gas modelled ΔNe also was significantly lower than observations. We identify that these discrepancies are due to deficiencies in two processes: (1) The model does not include a contribution from glacial meltwater (GMW) and (2) the surface cooling driven anomaly (ΔC_{surf}) appears to be too large in the model.

In the Southern Ocean, the addition of noble gases from glacial ice shelves contributes a source of dissolved noble gases not considered in our model. Glacial ice contains significant air content trapped in bubbles with a unique noble gas signature, as noble gas saturation anomalies are altered from atmospheric abundance by differential diffusion, gravitational enrichment and bubble processes [33]. Basal melting of glacial ice results in the addition of a distinct source of dissolved noble gases with the net effect of significantly increasing the saturation anomaly of the lighter gases and slightly decreasing

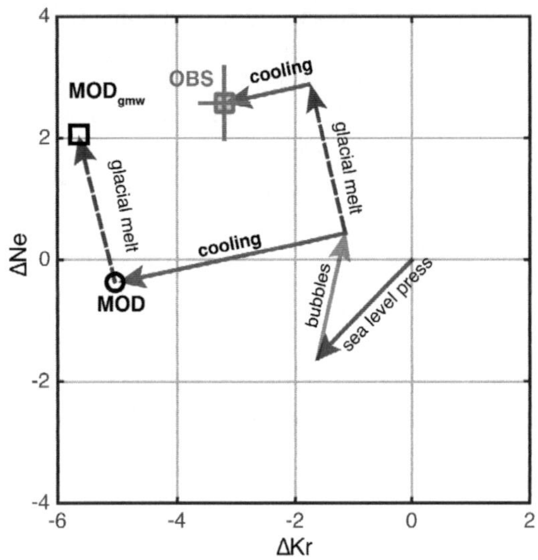

Figure 5. Model solution for the deep Weddell Sea (depth > 3500 m) is shown in Kr vs. Ne space. Results from ECCO (MOD) are shown with a black circle. After adding the glacial meltwater correction, modelled anomalies are located at the black square (MOD$_{gmw}$). Observations are indicated by the orange rectangle with 1 standard deviation error bars. The magnitude of cooling-driven saturation anomaly needed to reach MOD$_{gmw}$ is over twice that needed to reach OBS.

the saturation anomaly of heavier noble gases such that ΔNe_{gmw} = 8.9%, ΔAr_{gmw} = –0.5%, ΔKr_{gmw} = –2.2% and ΔXe_{gmw} = –3.3% per 1% volumetric contribution of GMW [10].

The cooling coupled with incomplete equilibration (the solubility pump) results in a saturation anomaly proportional to $D_C^{0.5}\left(1/[C]_{eq}\right)d[C]_{eq}/dT$ such that the heavy noble gases with slower diffusivity (D) and greater temperature dependence of solubility sustain a greater cooling-driven undersaturation (ΔC_{surf}). Thus, the sensitivity of Kr to cooling is much greater than that of Ne such that $\Delta Kr_{surf} / \Delta N_{surf} \approx 5.0$.

Equations for two gases, Ne and Kr, can be solved for the ΔC_{surf} and f_{gmw} needed to achieve the observed anomaly in the deep Weddell Sea (ΔNe = 2.6 ±0.6%, ΔKr = –3.2 ±0.4%) [10]. The solution is shown graphically in figure 5. Our solution yields an estimated f_{gmw} of 0.28 ± 0.07% in the deep Weddell Sea. Adding this correction factor to model results increases ΔNe by 2.4% and decreases ΔKr and ΔXe by 0.6% and 0.9%, respectively. The magnitude of cooling required to match observations was ΔNe_{surf} = –0.3% and ΔKr_{surf} = –1.5%. This observationally constrained cooling anomaly was much less than the cooling anomaly calculated directly by the model of ΔNe_{surf}= –0.8% and ΔKr_{surf} = –3.8%. Ne and Kr are an ideal pair for constraining these processes as the impact of these processes are nearly orthogonal in ΔKr vs. ΔNe space (figure 5). Similar results are obtained with other pairs of gases.

5. Discussion

The saturation anomaly in the abyssal deep Weddell Sea is composed of contributions from atmospheric pressure variation, air-sea bubble flux, glacial meltwater and the cooling-driven solubility pump. Here, we combine modelling and observations to estimate the contribution of glacial meltwater and the solubility pump. Noble gas constraints suggest a contribution of 0.28 ± 0.07% glacial meltwater to the deep Weddell Sea along the ANDREX section (figure 4).

Additionally, we identify a discrepancy between model and observations for the heavy noble gases that we attribute to an overly inefficient solubility pump in the ECCO-based model simulation. For the model the solubility pump component results in a ΔKr_{surf} = –3.8%, while observations suggest that ΔKr_{surf} = –1.5% is more realistic (figure 5).

From equation (6) we see that in addition to physical properties of the noble gases, the magnitude of cooling-driven saturation anomaly scales proportionally with $-H/k(1-f_{ice})$. Thus we consider air-sea heat flux and the diffusive gas exchange rate as the likely sources of deficiency in the model simulations. A cooling flux of too great magnitude could be indicative of the tendency of 1° global models to create Southern Ocean deep water via open ocean convection, rather than shelf processes [34]. The regions where the deep Weddell Gyre was ventilated are shown in supplementary figure S2. Given the relatively high ice-cover in the Weddell Sea, a bias in the ice inhibition of air-sea transfer term, $(1-f_{ice})$, could have a large impact on the solubility pump as well. We note that sea ice and abyssal water formation through density currents are both particularly challenging problems for most earth system models including ECCO [34–36].

Whatever the cause, insufficient ventilation at the time and location of deepwater formation will have large implications for the global biogeochemical cycles of carbon and a range of dissolved gases. Noble gas tracers indicate that model deepwater retains a cooling-induced saturation anomaly that is over twice what observations suggest it should be. Such model shortcomings might be expected to result in underestimation of the uptake of anthropogenic carbon and transient tracers in the Southern Ocean. Our analysis suggests the promise of noble gas tracers as an independent way of assessing the fidelity of ocean circulation models (and state estimates). The future availability of TMM matrices for a range of models and model configurations will enable an efficient framework for such an assessment.

References

[1] Broecker W S and Peng T 1982 *Tracers in the Sea* (Palisades, New York: Lamont-Doherty Geological Observatory, Columbia University)
[2] Toggweiler J R and Sarmiento J L 1985 Glacial to interglacial changes in atmospheric carbon dioxide: The critical role of ocean surface water in high latitudes *The Carbon Cycle and Atmospheric CO: Natural Variations Archean to Present* eds E T Sundquist and W S Broecker (Washington, DC: American Geophysical Union)
[3] Toggweiler J R *et al* 2003 *Glob. Biogeochem. Cycles* **17** 1026
[4] Maier-Reimer E and Hasselmann K 1987 *Clim. Dyn.* **2** 63
[5] Jones D C *et al* 2014 *Glob. Biogeochem. Cycles* **28** 1163
[6] Marinov I *et al* 2006 *Nature* **441** 964
[7] Ferrari R *et al* 2014 *Proc. Natl. Acad. Sci.* **111** 8753
[8] Hamme R C and Severinghaus J P 2007 *Deep-Sea Res. Part I* **54** 939
[9] Nicholson D *et al* 2010 *J. Geophys. Res. Oceans* **115** C11015
[10] Loose B and Jenkins W J 2014 *Geophys. Res. Lett.* **41** 2013GL058804
[11] Stanley R H R and Jenkins W J 2013 Noble Gases in seawater as tracers for physical and biogeochemical ocean processes *The Noble Gases as Geochemical Tracers* Advances in Isotope Geochemistry ed P Burnard (Springer Berlin Heidelberg) 55–79
[12] Stanley R H R *et al* 2009 *J. Geophys. Res. Oceans* **114** C11020
[13] Nicholson D, *et al* 2011 An inverse approach to estimate bubble-mediated air-sea gas flux from inert gas measurements *Proceeding of the 6th international symposium on gas transfer at water surfaces* (Kyoto, Japan: Kyoto University Press)
[14] Ito T and Deutsch C 2006 *Glob. Biogeochem. Cycles* **20** GB3019
[15] Emerson S *et al* 2012 *Geophys. Res. Lett.* **39** L18610
[16] Hamme R C and Emerson S R 2004 *Deep-Sea Res. Part I* **51** 1517
[17] Weiss R F and Kyser T K 1978 *J. Chem. Eng. Data* **23** 69
[18] Wood D and Caputi R 1966 *Solubilities of Kr and Xe in fresh and sea water* USNRDL TR 988 (San Francisco, CA: US Naval Radiological Defense Laboratory)
[19] Keeling R F 1993 *J. Mar. Res.* **51** 237
[20] Fuchs G *et al* 1987 *J. Geophys. Res.-Oceans* **92** 6559
[21] Jenkins W J *et al* 1988 *Philos. Trans. R. Soc. Lond. Ser. Math. Phys. Sci.* **325** 43
[22] Hamme R C and Emerson S R 2006 *J. Mar. Res.* **64** 73
[23] Wunsch C *et al* 2009 *Oceanography* **22** 88
[24] Adcroft A *et al* 2004 Overview of the formulation and numerics of the MIT GCM *Proceedings of the ECMWF Seminar on "Recent developments in numerical methods for atmospheric*

and ocean modelling" (Shinfield Park, Reading, UK) 139–50
[25] Wunsch C and Heimbach P 2007 *Phys. Nonlinear Phenom.* **230** 197
[26] Wunsch C and Heimbach P 2013 Dynamically and kinematically consistent global ocean circulation and ice state estimates Ocean Circulation and Climate: A 21st Century Perspective eds G Siedler *et al* (Academic Press) 553–79
[27] Khatiwala S *et al* 2005 *Ocean Model.* **9** 51
[28] Khatiwala S 2007 *Glob. Biogeochem. Cycles* **21** doi:10.1029/2007GB002923
[29] Khatiwala S 2008 *Ocean Model.* **23** 121
[30] Sweeney C *et al* 2007 *Glob. Biogeochem. Cycles* **21** doi:10.1029/2006GB002784
[31] Large W G and Yeager S G 2008 *Clim. Dyn.* **33** 341
[32] Jähne B *et al* 1987 *J. Geophys. Res.* **92** 10767
[33] Severinghaus J P and Battle M O 2006 *Earth Planet. Sci. Lett.* **244** 474
[34] Heuzé C *et al* 2013 *Geophys. Res. Lett.* **40** 1409
[35] Legg S *et al* 2009 *Bull. Am. Meteorol. Soc.* **90** 657
[36] Turner J *et al* 2012 *J. Clim.* **26** 1473

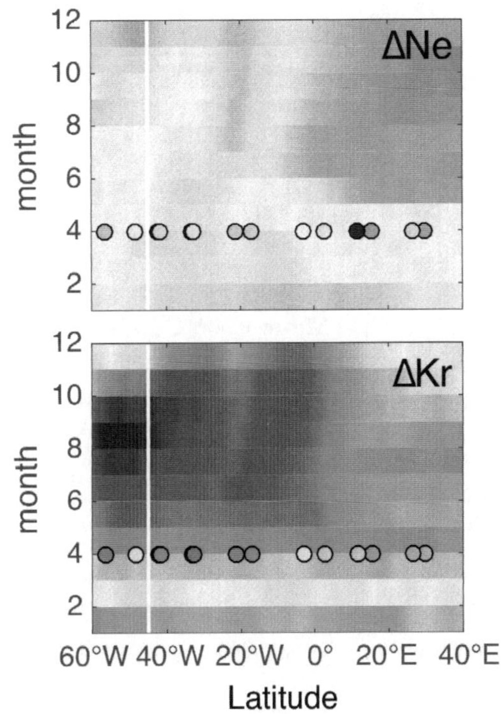

Figure S1. Hovmöller diagram of monthly mean surface ΔNe and ΔKr along 60°S with ANDREX observations from April 2010 overlaid.

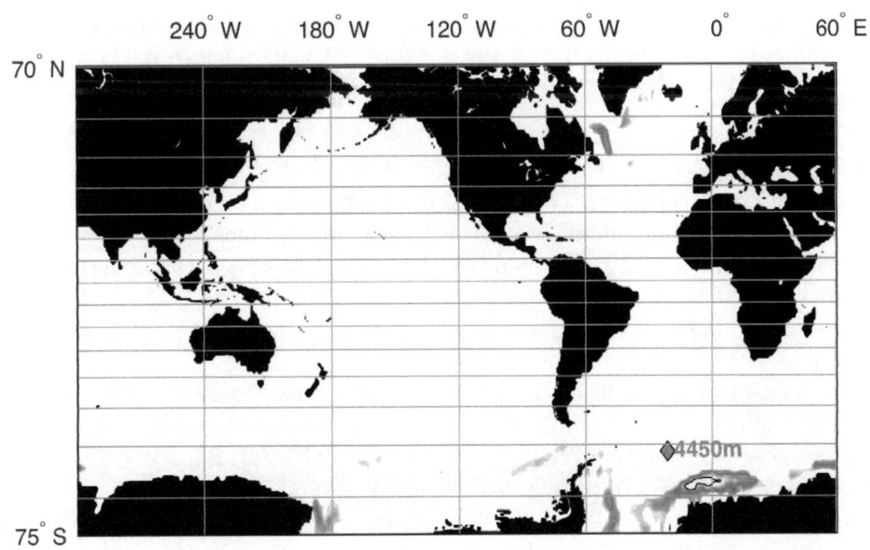

Figure S2. Results for an adjoint Green function calculation for a target volume in the deep Weddell Gyre (21.3°W 61.6°S at 4450 m) was last ventilated. The pink marker indicates the location of the deep (4450 m) target volume. The solid black contour encloses a region accounting for where 50% of the water parcel was ventilated and the light grey contour encloses where 90% of ventilation occurred.

Gas transfer in a bubbly wake flow

A Karn[1,2], J S Gulliver[1,3], G M Monson[4], C Ellis[1], R E A Arndt[1] and J Hong[1,2]

[1] Saint Anthony Falls Laboratory, University of Minnesota, Minneapolis, MN, USA
[2] Department of Mechanical Engineering, University of Minnesota, Minneapolis, MN, USA
[3] Department of Civil, Environmental and Geo Engineering, University of Minnesota, Minneapolis, MN, USA
[4] Houston Engineering, Maple Grove, MN, USA

E-mail: gulli003@umn.edu

Abstract. The present work reports simultaneous bubble size and gas transfer measurements in a bubbly wake flow of a hydrofoil, designed to be similar to a hydroturbine blade. Bubble size was measured by a shadow imaging technique and found to have a Sauter mean diameter of 0.9 mm for a reference case. A lower gas flow rate, greater liquid velocities, and a larger angle of attack all resulted in an increased number of small size bubbles and a reduced weighted mean bubble size. Bubble-water gas transfer is measured by the disturbed equilibrium technique. The gas transfer model of Azbel (1981) is utilized to characterize the liquid film coefficient for gas transfer, with one scaling coefficient to reflect the fact that characteristic turbulent velocity is replaced by cross-sectional mean velocity. The coefficient was found to stay constant at a particular hydrofoil configuration while it varied within a narrow range of 0.52–0.60 for different gas/water flow conditions.

1. Introduction

Air injection has often been suggested as a method to improve DO concentration in lakes and rivers. Air injected into the water in ventilation systems such as vented turbines breaks up into small bubbles [1]. In the downstream of these turbines, a bubbly wake flow is created where the turbulence levels are high. In such a bubbly wake flow, the velocity fluctuations and turbulence in the flow breaks up larger bubbles into smaller size bubbles, thus increasing the total gas-water interfacial area and augmenting gas transfer [2, 3].

In this study, a series of air injection experiments were conducted on a hydrofoil (turbine blade prototype design) at several hydrodynamic conditions to investigate the effects of flow field on the size of the bubbles generated and the resulting rate of oxygen transfer. Experiments were designed to study the impact of varying water velocities, airflow, and hydrofoil orientation (angle of attack) on the bubble sizes generated, void fraction in the wake, and oxygen transfer from the bubbles. Further, a one-dimensional mass transfer model is developed to relate the dependence of the mass transfer characteristics with the bubble sizes observed in our experiments.

2. Experimental methodology

The experiments were conducted in the Saint Anthony Falls Laboratory (SAFL) high-speed water tunnel at the University of Minnesota. As shown in figure 1, the tunnel has a horizontal test section of 1.2 m (length) × 0.19 m (width) × 0.19 m (height) with three side walls made of plexiglass for optical

Content from this work may be used under the terms of the Creative Commons Attribution 3.0 licence. Any further distribution of this work must maintain attribution to the author(s) and the title of the work, journal citation and DOI.
Published under licence by IOP Publishing Ltd

access. The tunnel is designed for cavitation and air ventilation studies and is capable of reaching water velocities as high as 20 m/s. The gas collector dome (in the settling chamber) of the tunnel provides for the removal of large quantities of air injected as part of ventilation experiments, thus allowing such experiments to continue for extended periods of time without affecting test section conditions.

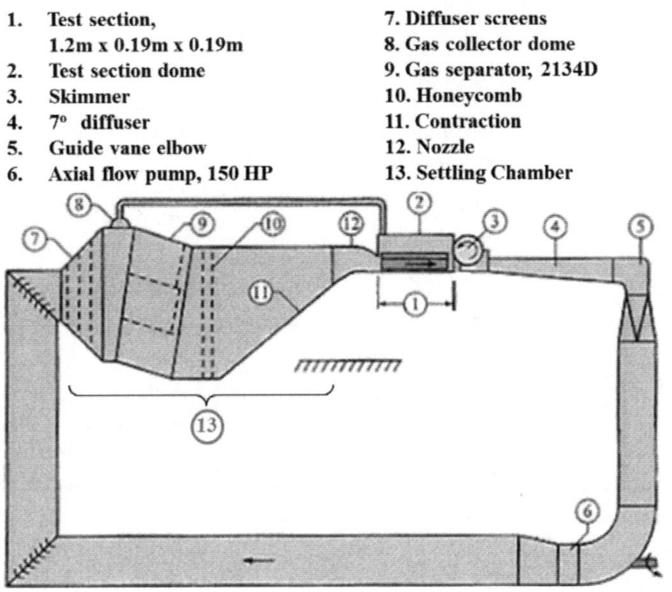

Figure 1. Schematic of the SAFL water tunnel facility.

In our experiments, a two-dimensional NACA0015 hydrofoil was installed horizontally in the test section and adjusted to one of three different angles of attack (α): 0°, 4° and 8°. The hydrofoil measures 190 mm in span and 81 mm in chord. As figure 2 shows, a narrow spanwise slot allowed air to be injected into the flow near the leading edge of the hydrofoil on its suction side. For the gas transfer measurements, the full span length of the injection slot was used for air injection. This results in a dense spanwise bubbly wake. However, in order to make bubble measurements, ventilation was limited to a narrow 9.6-mm long slot section (5% of the full slot length) at the centre of the span. This configuration ensures that bubbles remain within a narrow depth of field from the centre of the test section. Videos of the bubbly wake verify that the lateral drift or spread of the bubbles was minimal. Fifteen different experiments were conducted at test section water speeds of 5, 7.5 and 10 m/s, ventilation gas flow of 0.5, 1.0 and 1.5 SLPM, and hydrofoil angle of attack (AoA) of 0°, 4° and 8°. The bubbly wake images were obtained at three streamwise distances of 109, 243 and 377 mm downstream of the hydrofoil centre (or 1.3, 3.0 and 4.7 times the chord length, respectively.) The resulting bubble diameters were close to the same at the three distances downstream, and the 377-mm location was chosen because it was believed to yield the best view of the bubbles for bubble size measurement.

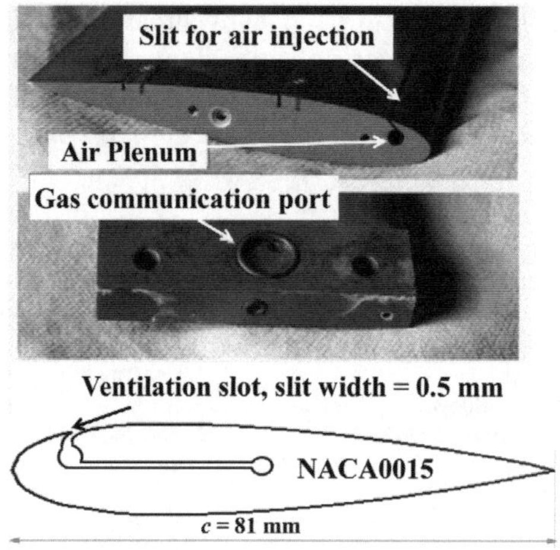

Figure 2. Details of the NACA0015 ventilated foil.

2.1 Experiment methodology: Degassing and imaging

A disturbed equilibrium technique was used to estimate gas transfer rates by initially degassing the water to the lowest level possible and then re-oxygenating the water while taking continuous DO concentration measurements [4]. During degassing, a portion of water is withdrawn from the tunnel and circulated through a degassing chamber under vacuum where the water is discharged into the chamber through nozzles, creating a cloud of falling droplets. The degassing loop is run overnight, reducing the DO concentration to approximately 4 ppm. Immediately prior to an aeration test, the water tunnel is run without air injection until the water is thoroughly mixed. A test commences by setting the test section velocity and airflow rate. The experiment is run for approximately 10 hr or until the oxygen concentration achieves 99% of steady state.

A shadow imaging technique [1–4] with direct, in-line volume illumination using a pulsed LED array source was used to capture shadow images of the bubbly wake similar to the study of Karn *et al.* [1]. Figure 3 shows a schematic of the experimental setup. A custom-made pulsed LED light source from Innovative Scientific Solutions, Inc. was used to illuminate the flow. The pulsed LED array has flash rates up to 10 kHz with a 5-μs minimum pulse width and rise and fall times of ~ 200 ns. To improve the uniformity of the back-lighting in the images, a light shaping diffuser was placed between the light source and the flow. A 1K × 1K pixel Photron APX-RS camera, capable of 3000 frames/s at full resolution was used to obtain images. To minimize blur, all exposure times were 15 μs and the frame rate was 25 images per second. The focus plane was located at the centre of the test section and the depth of field was determined to be approximately 10 mm. The captured images have a spatial resolution of 0.059 mm/pixel and the bubble sizes are in the range of 1 to 68 pixels, which corresponds to a bubble diameter of 0.12 to 4 mm. A data set for a single experiment consists of 1000 images taken over a 40-s period.

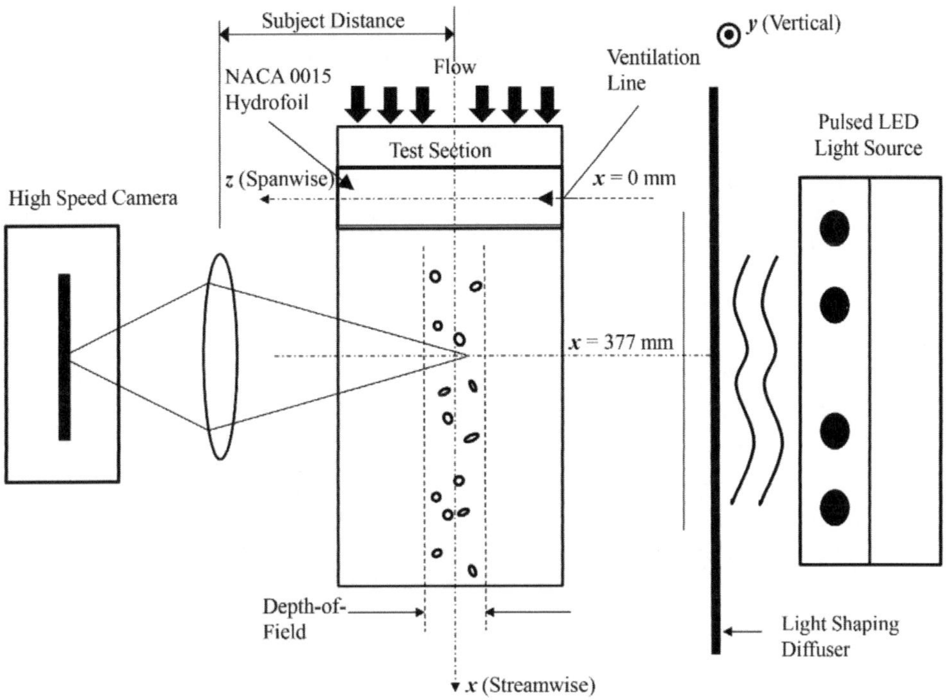

Figure 3. Schematic of the experimental setup for shadow imaging of bubbly flows.

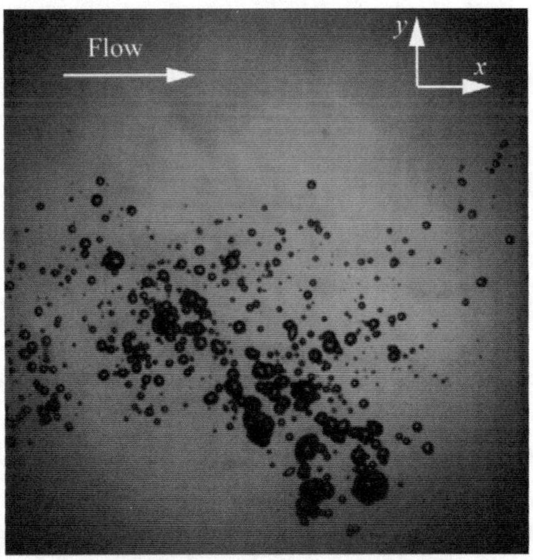

Figure 4. A sample bubbly wake image obtained in shadow imaging experiments.

2.2 *Image processing and analysis*

Figure 4 provides a sample bubble image from our experiments. The bubbly wake is characterized by a wide bubble size distribution ranging from microbubbles to large bubble clusters. The bubbles could be either in-focus or slightly out-of-focus. An image analysis technique, as described in Karn *et al.* [5], is employed to extract the information about the bubble locations and the bubble sizes from the bubble images. It consists of three major steps. (1) The original grayscale images are first binarized. This is followed by (2) categorization of bubble regions. The bubble regions are labeled and characterized by a series of metrics including area, centroid, major and minor radius (projected bubbles are assumed to be ellipses) and a circularity factor. Based on the area of each region, the bubbles are divided into tiny spherical bubbles, intermediate-sized bubbles and large bubbles or clusters. Last (3), bubble separation and information extraction is performed. A multilevel segmentation approach is used to extract the bubble locations and dimensions from the images. Cluster processing using advanced morphological operations and a watershed transform extracts individual bubbles from bubble clusters. The information for all identified bubbles, including their x and y-coordinates and minor and major radius is then tabulated.

3. Gas transfer model

A one-dimensional model of the change in dissolved oxygen (DO) concentration in the tunnel was developed to fit the results of the physical experiments to a mass transfer theory using a single coefficient, as in Karn *et al.* [4]. The simplified conservation equation for the concentration of oxygen in a section of the tunnel is given by equation 1 as:

$$U\frac{dC}{dx^*} = k_L a\,(C_s - C) \tag{1}$$

where C refers to DO concentration and C_S corresponds to equilibrium DO concentration in a particular section of the tunnel, x^* refers to a moving co-ordinate system given as $x^* = x - Ut$, U is cross-sectional mean velocity, k_L is the liquid film coefficient for transfer of a volatile gas across the bubble-liquid interface, and a is the specific surface area of bubbles (or the surface area of bubbles divided by volume associated with those bubbles), given by equation 2 below:

$$a = \frac{A_{St}}{V_w} \tag{2}$$

where A_{St} is the total surface area of the bubbles and V_w is the volume of water in which A_{St} is determined. Bubble diameter is determined from the shadow imaging experiments, with limited width, and surface area weighted bubble diameter is defined as:

$$d_2 = \sqrt{\frac{A_{si}}{\pi N_{si}}} \tag{3}$$

where A_{si} and N_{si} are the bubble surface area and the number of bubbles from the shadow imaging experiments, respectively. Next, it is assumed that the limited width of the shadow imaging experiments does not alter this ratio:

$$\frac{A_{si}}{N_{si}} = \frac{A_{St}}{N_t} \tag{4}$$

where N_t is the total number of bubbles corresponding to A_{St}. Then, the combination of equations 3 and 4 gives:

$$A_{St} = \pi d_2^2 N_t. \tag{5}$$

The void fraction of air (φ) is given as:

$$\varphi = \frac{V_b}{V_w} = \frac{N_t \pi d_3^2}{6V_w} \tag{6}$$

where V_b is the total volume of bubbles in control volume V_w, and d_3 is the volume weighted bubble diameter determined from the shadow imaging experiments:

$$d_3 = \left(\frac{6V_{si}}{\pi N_{si}}\right)^{1/3} \tag{7}$$

where V_{si} is the total volume of bubbles from the shadow imaging experiments, and equation 6 may be solved for N_t:

$$N_t = \frac{6\varphi V_w}{\pi d_3^3} \tag{8}$$

Equations 2, 5 and 8 may be combined to give:

$$a = 6\varphi\,\frac{d_2^2}{d_3^3} = \frac{6\varphi}{d_{32}} \tag{9}$$

where d_{32} is the Sauter mean diameter of the generated bubble population. Thus, Sauter mean diameter is the appropriate weighted diameter to use in gas transfer calculations.

The equilbrium concentration C_s is given by Henry's law and varies with temperature, pressure and solute content. The gas transfer coefficient k_L depends on the local air void fraction, the geometrical characteristics of the bubbles and local flow characteristics. Therefore, k_L may be approximated by predictive relations [6, 7], but is best determined by controlled, flow-specific laboratory experiments. The theory of Azbel [8] is employed to determine the liquid film coefficient in the model, k_L as:

$$k_L = \beta \frac{D^{1/2}}{4\pi} \frac{u^{3/4}}{(v_f l)^{1/4}} \frac{(1-\varphi)^{1/2}}{(1-\varphi^{5/3})^{1/4}} \tag{10}$$

where D is the diffusion coefficient of the solute (oxygen) in water, v_f is kinematic viscosity, l is a characteristic turbulent length scale, u is a characteristic turbulent velocity, and β is a coefficient on the order of 1 to be determined by fitting with the experimental data. Note that $u^{3/4}/\left(v_f l\right)^{1/4}$ is equal to the inverse square root of a time constant of the flow field.

4. Results

Figure 5 shows the probability density functions (PDF) of number of bubbles, total surface area and total volume as functions of bubble diameter in the wake of the foil for zero degrees angle of attack, a water velocity of 5 m/s and a ventilation mass airflow rate of 10 SLPM (standard liters per minute). The PDF of bubble number illustrates the large fraction of very small size bubbles in the wake. As can be seen from the plot, more than 40% of the bubbles are between 0.2 and 0.4 mm in diameter. Karn *et al.* [2] suggest that these small size bubbles are not generated at the air injection point but are produced as a result of breakup processes related to increased turbulence in the wake. The bubbles below 0.4 mm only produce 5% of the total bubble surface area, and thus are not a substantial contributor to gas transfer.

The PDFs of the bubble surface area and bubble volume in figure 5 exhibit maxima just below and just above a bubble diameter of 1 mm, respectively. Around 24% of the surface area of the bubbles is contributed by bubbles with a diameter of between 0.8 mm and 1 mm. An equal amount of total bubble surface area is represented by slightly larger bubbles with diameters 1 – 1.2 mm.

Table 1 shows the variation of d_{32} across all conditions. A tripling of ventilation flow rate produces an approximately 14% increase in d_{32} at both water velocities and three angles of attack. Similarly, an increase between 44% and 55% in d_{32} is observed when water velocity is halved at both airflow rates, and 4° and 8° angles of attack. The increase is somewhat smaller (31% and 33%) at zero angle of attack. It can also be seen from table 1 that there is a slight decrease in d_{32} with increasing angles of attack, and a consistent trend can be observed. These observations demonstrate the importance of liquid velocity and turbulence in determining bubble size, as the bubbles are sheared off from entry and split multiple times by the associated forces.

Table 1 also shows the variation of the β coefficient over the range of experimental conditions tested. It is seen that, except for the β value of 0.34 at AoA = 8°, U = 10 m/s, air flow = 10 SLPM, which is considered an outlier, the value of β depends only on the hydrofoil configuration and has close to a constant value for all the flow conditions at a particular AoA. The β coefficient increases from approximately 0.53 to 0.60 as AoA is varied from 0 to 8°. The fact that β is less than its expected value of one could be due to the cross-sectional mean velocity being substituted for the characteristic turbulent velocity in equation 10. If the characteristic turbulent velocity is approximately 0.43 (AoA = 8°) to 0.51 (AoA = 0°) times the free stream mean velocity, the value of β, as calculated from equation 10, would be 1. A value of 0.43 to 0.51 for the ratio of the characteristic turbulent velocity to mean velocity is realistic for the flow in the wake of a hydrofoil. Determining a β value in the range of 0.53–0.60 for different hydrofoil angles of attack, independent of water velocity and airflow rate, is seen as a confirmation of Azbel's [8] theory.

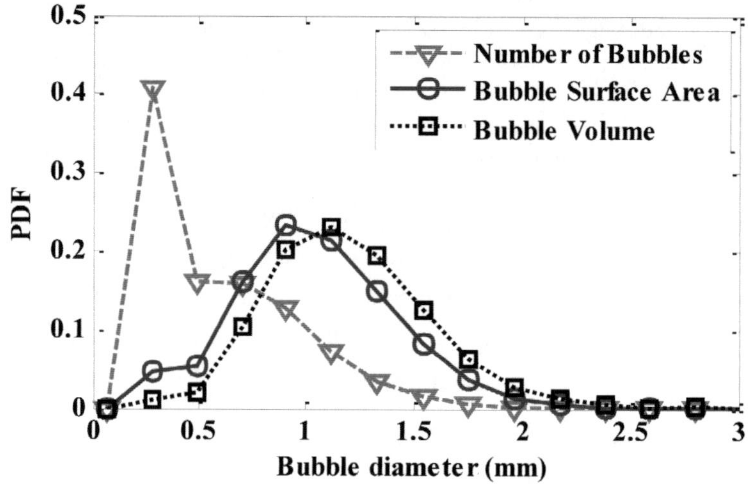

Figure 5. PDFs of bubble numbers, bubble surface area and volume for the reference case.

AoA (deg)	U (m/s)	Q (SLPM)	d_{32} (mm)	β
8	5	10	0.79	0.54
8	5	30	0.89	0.54
8	7.5	20	0.68	0.54
8	10	10	0.51	0.34
8	10	30	0.58	0.52
4	5	10	0.84	0.54
4	5	30	0.95	0.59
4	7.5	20	0.75	0.59
4	10	10	0.58	0.59
4	10	30	0.66	0.54
0	5	10	0.88	0.60
0	5	30	0.98	0.60
0	7.5	20	0.80	0.60
0	10	10	0.66	0.60
0	10	30	0.75	0.60

Table 1. Description of gas transfer experiments conducted in the present study with all the relevant experimental conditions. AoA is the angle of attack. d_{32} is obtained from images. The β values are determined from a fit of the gas transfer model to experiments.

5. Conclusions

Experiments on bubble-water mass transfer were performed in a closed, recirculating high-speed water tunnel using a ventilated hydrofoil (designed to be similar to a hydroturbine blade) by varying the free stream water velocity, ventilation air flow rate and the hydrofoil angle of attack. The current study shows the effect of these parameters on the size of the bubbles generated by air injection and the resulting rate of oxygen transfer in the wake of a hydroturbine. The shadow imaging experiments in the bubbly wake revealed that for our reference case of 0° AoA, a water velocity of 5 m/s and a ventilation mass airflow rate of 10 SLPM, ~40% of the bubbles in the wake had diameters of less than

0.4 mm. It is also found that bubbles larger than 0.4 mm in diameter have a greater influence on the Sauter mean diameter of the population.

A one-dimensional mass transfer model is developed to lend insight into the oxygen transfer characteristics observed in these experiments using a mass transfer conservation equation and a predictive relation for liquid film coefficient, k_L, by Azbel [8]. For ease of application, mean free stream velocity (U) was substituted for characteristic turbulent velocity (u') in the predictive relation. Thus, a β coefficient was introduced to account for this difference, $\beta = (u'/U)^{3/4}$, where u'/U is the turbulence intensity. This β coefficient was determined by fitting the prediction of the numerical model with the experimental data obtained from the DO measurements in the water tunnel. The gas transfer model required the test conditions and Sauter mean diameter of the bubbles in the wake of the hydrofoil as input to predict the liquid film coefficient, k_L. Except for one outlier, the results of the model showed that β coefficient is constant for a given angle of attack. β decreases slightly as hydrofoil AoA increases. We believe the values of β derived in this study can be used with Azbel's k_L relationship as a predictive tool for mass transfer in bubbly flows, which are similar to those studied here. Prior studies have largely been unsuccessful due to the lack of experimental measurements of bubble size distributions, forcing a reliance on empirical formulations for bubble size and surface area [6, 7]. It is hoped that these experimental results on bubble size and mass transfer will provide further insight into the design of hydroturbine blades.

References

[1] Karn A *et al* 2015 *Exp. Therm. Fluids Sci.* **64** 186
[2] Karn A *et al* 2015 *Exp. Therm. Fluids Sci.* **70** 397
[3] Karn A *et al* 2015 *Int. J. Fluid Machinery Syst.* **8** 73
[4] Karn A *et al* 2015 *Int. J. Heat Mass Trans.* **87** 512
[5] Karn A *et al* 2015 *Chem. Eng. Sci.* **122** 240
[6] Wilhelms S C and Gulliver J S 2002 *J. Hydraul. Res.* **43** 532
[7] Witt,A M and Gulliver J S 2012 *J. Hydraul. Res.* **50** 521
[8] Azbel D 1981 *Two-phase Flows in Chemical Engineering* ed Kemp-Pritchard P (New York: Cambridge University Press)

IOP Conf. Series: Earth and Environmental Science **35** (2016) 012021 doi:10.1088/1755-1315/35/1/012021

Development of thermal image velocimetry techniques to measure the water surface velocity

A Saket[1], W L Peirson[1], M L Banner[2] and X Barthelemy[1,2]
[1] Water Research Laboratory, School of Civil and Environmental Engineering, UNSW Australia, 110 King Street, Manly Vale, NSW 2093, Australia
[2] School of Mathematics and Statistics, UNSW Australia, Sydney, NSW 2052, Australia

Email: a.saket@wrl.unsw.edu.au, w.peirson@unsw.edu.au, m.banner@unsw.edu.au, and x.barthelemy@wrl.unsw.edu.au

Abstract. Particle image velocimetry (PIV) is a state-of-the-art non-intrusive technique for velocity and fluid flow measurements. Due to ongoing improvements in image hardware and processing techniques, the diversity of applications of the PIV method continues to increase. This study presents an accurate thermal image velocimetry (TIV) technique using a CO_2 laser source to measure the surface wave particle velocity using infrared imagery. Experiments were carried out in a 2-D wind wave flume with glass side walls for deep-water monochromatic and group waves. It was shown that the TIV technique is robust for both unforced and wind-forced group wave studies. Surface wave particles attain their highest velocity at the group crest maximum and slow down thereafter. As previously observed, each wave crest slows down as it approaches its crest maximum but this study demonstrates that the minimum crest speed coincides with maximum water velocity at the wave crest. Present results indicate that breaking is initiated once the water surface particle velocity at the wave crest exceeds a set proportion of the velocity of the slowing crest as it passes through the maximum of a wave group.

1. Introduction

Visualisation techniques play a central role in fluid mechanics in quantifying the behaviour of a flow field. One of the main non-intrusive techniques for velocity and fluid flow measurements is particle image velocimetry (PIV). Due to ongoing improvements in image hardware and processing techniques, the diversity of applications of the PIV method continues to increase. Advances in PIV techniques have led to experimental investigations of complex problems providing a foundation for development of more accurate numerical models. During the last two decades, the PIV technique has been widely used in the field of wave mechanics [1–7].

Understanding of velocities and accelerations of waves is essential for calculating forces on marine structures. For breaking waves this force is potentially much larger than for non-breaking waves. Therefore, reliable characterisation of the wave hydrodynamics particularly for breaking waves is important to improve the design of ocean and coastal structures.

A fundamental and long-standing gap in our understanding of deep-water wave breaking is how to characterise and predict the breaking onset. The kinematic breaking criterion based on surface particle and wave crest velocities is often suggested to predict the onset of breaking. The surface kinematics

Content from this work may be used under the terms of the Creative Commons Attribution 3.0 licence. Any further distribution of this work must maintain attribution to the author(s) and the title of the work, journal citation and DOI.
Published under licence by IOP Publishing Ltd

associated with wave breaking has been studied using a suite of imaging techniques during the last two decades [1, 8–13]. Because measuring the surface particle velocity and the defining the crest speed are difficult, evaluation of this criterion remains nontrivial.

Banner *et al.* [14] studied deep water wave kinematics numerically, both in the laboratory and the field and have shown that the dominant waves in a group systematically slow down as they pass through the group maximum by between 10 and 30%. Barthelemy *et al.* [15] show numerically that the onset of breaking occurs once the water surface particle velocity at the wave crest exceeds a set proportion of the velocity of the slowing crest as it passes through the maximum of a wave group.

This contribution presents evaluation of a thermal image velocimetry (TIV) technique to measure the surface wave velocity using infrared imagery. Experiments were carried out in a 2D wave flume with glass side walls. Hot water droplets and a CO2 laser were used to create localised thermal signatures on the water surface, which were tracked and analysed.

2. The experiment

2.1. Experimental facilities

The experiments were conducted in the two-dimensional wind-wave tank at the Water Research Laboratory used previously by Banner and Peirson [16] (figure 1). The wave tank is 30 m long, 0.6 m wide and 0.6 m high with glass side walls and was filled with water to the depth of 0.46 m. Waves were generated using a computer-controlled, flexible bottom-cantilevered wave paddle located at one end of the flume. At the other end, a flexible reticulated polyester-urethane foam absorbent beach was installed to minimise reflections of the generated waves.

A movable wind tunnel with the length of 7.5 m and the roof of 0.5 m above the still water surface was mounted on the flume. A fan installed at the downstream of the tunnel generated wind flow and an adjustable honeycomb flow provided a uniform air flow within the wind tunnel. The wind speed was measured at the centre of the tunnel approximately 4.8 m downwind of the inlet and 0.25 m above the still water level using a pre-calibrated hot probe air velocity meter (Velocicalc model 8347).

For generating wind-forced waves, wind intensity was controlled by varying the fan input voltage using a metered Variac. Paddle-generated waves under wind forcing were investigated for different wind speeds in the range of 1.0 to 2.2 m/s.

To find the paddle amplitude corresponding to the transition between maximum (non-breaking) recurrence and marginal breaking, light was projected inside the tunnel and the transitional condition was determined visually. The paddle amplitude was increased systematically until the location of initial breaking was detected.

Two linear arrays of capacitance wave probes with 200 mm wire elements were located at each side of the flume to record the water elevation using a National Instruments PCI-6225 data acquisition system at 1000 Hz sample rate per channel. The probe resolution was 0.1 mm with a linearity of ±0.2 mm over their 200 mm length. To minimise any effect of the probes on the wind flow, the wave probe signal conditioning boxes were mounted outside the tunnel, with probes connected to the boxes via 6 mm diameter cables.

Five wave probes were installed along the one side of the tank with the spacing of 60 mm to measure the crest speed. At the other side of the tank, seven wave probes with the spacing of 100 mm were installed to measure the wave length. The water level time series captured by the wave probes were interpolated in space to obtain the zero-crossing locations at the time of the wave group maximum and, thereby, the length of the crest in the direction of propagation and crest elevation. The local steepness S_c was calculated using the length of the crest in the direction of propagation and the maximum crest elevation for each wave group.

2.2. Wave conditions

To measure the wave crest speed and water particle velocity, monochromatic and group waves were generated by the paddle. The initial conditions for generating wave packets were based on Banner and Peirson [2007].

Bimodal spectrum (Class 2) and chirped (Class 3) initial wave packets were investigated in the current study. The form of the bimodal initial spectrum was taken as:

$$\eta = a_0 \cos(k_0 x) + \varepsilon a_0 \cos\left(\frac{N+1}{N} k_0 x - \frac{\pi}{18}\right) \tag{1}$$

where η is the water surface elevation, a_0 is the initial amplitude, k_0 is the wave number, N is the number of waves in the group and ε=0.1.

The chirped wave packet was taken to have the form:

$$x_p = a - 0.25 A_p \left(1 + \tanh \frac{4\omega_p t}{N\pi}\right)\left(1 - \tanh \frac{4(\omega_p t - 2N\pi)}{N\pi}\right) \sin\left(\omega_p (t - \omega_p C_{t2} t^2 / 2)\right) \tag{2}$$

where x_p represents the wave paddle motion, A_p is proportional to the piston amplitude, t is time, ω_p is the paddle angular frequency and C_{t2} is the chirp rate of the linear modulation.

3. TIV system

In this study, two TIV methods were used to measure the horizontal water particle velocities. The system consisted of heat sources, an infrared imaging camera, a computer-controlled shutter, an acoustic sensor, synchroniser and a computer controlling the system components.

3.1. Localised thermal signatures on the water surface

Initially a thermos flask was used to deposit hot water droplets on the water surface with low (<10) Weber number ($W = \varrho u^2 d/\sigma$), where ϱ is the droplet density, u is the mean droplet impact velocity, d is the droplet diameter and σ is the surface tension. The thermos was mounted at the centre line of the flume with two linear arrays of wave probes just inside each of the tank side walls on a trolley (figure 1a). To adjust the time interval between droplets and the dimension of the droplets, a valve and nozzle were installed on the thermos. For group waves, hot droplets were laid down upstream of the group maximum location.

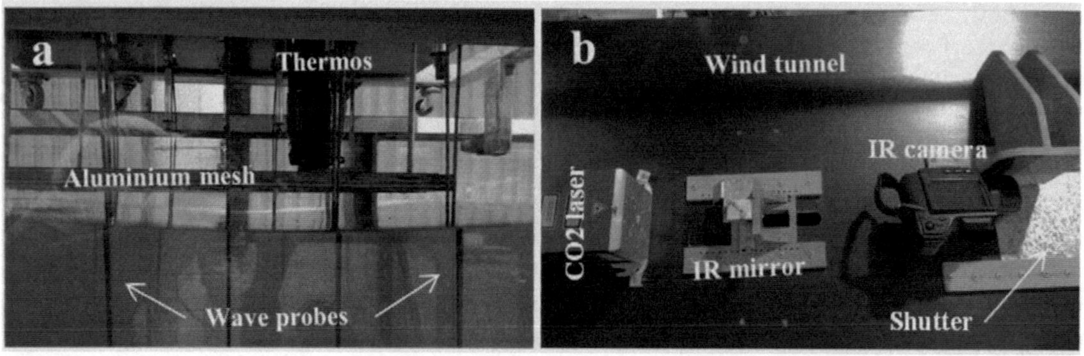

Figure 1. Laboratory deployment of instrumentation: (a) Side view showing positioning of the Thermos, wave probes and an aluminium calibration mesh installed on a trolley. (b) View from above showing the wind tunnel, CO_2 laser system, adjustable IR mirror, infrared imaging camera and computer-controlled shutter. Note that the Thermos and calibration mesh were removed once the laser system was deployed.

In order to improve the localisation of hot spot formation on the water surface, a CO_2 laser (Firestar ti100) system was developed to irradiate the water surface. The laser was mounted at the centre of the wind tunnel roof to produce heated circular patches with a diameter of approximately 4 mm upstream of the group maximum location (figure 1b).

As CO_2 lasers produce a beam of infrared light with 10.6 μm wavelength, an IR flat mirror was installed on an adjustable holder in front of the laser head to reflect the laser light on to the water surface. To minimise the diameter of the heated patches, the duty cycles were set as small as possible. The range of laser duty cycles for different wave packets was selected between 5 to 10 ms.

3.2. Infrared imaging camera

The TIV system used an infrared imaging camera (Flir T420) to acquire images of surface displacements from above, capturing 320×240 pixel images of the surface with the 25°×19° field of view at 30 frames per second (figure 1b). The physical resolution of the thermal imager was approximately 0.66 mm per pixel at the water surface, determined using calibration grids placed within the field of view.

For the TIV system using hot water, the camera was mounted on the trolley at the centre line of the tank and central to the wave probe arrays. An aluminium mesh was installed in the field of view of the camera. To adjust the level of the grids for calibrating the images, the mesh frame was connected to the trolley using four adjustable bars. In the CO_2 system, the thermal camera was mounted at the centre of the wave probe arrays on the wind tunnel roof to observe the tank water surface through a shuttered window 3.6 m downwind of the wind tunnel inlet. Because the laser reflections could potentially damage the camera's image sensor, a computer-controlled shutter was installed beneath the camera and the shutter remained closed during surface irradiation.

3.3. Synchronisation

In the TIV system, one of the wave probes was applied as a trigger to synchronise the wave signals, laser pulses and the shutter using a National Instrument PCI-6221 data acquisition system. At a selected water surface level, laser pulsing commenced with a defined duty cycle. After surface irradiation, the shutter was opened and the thermal camera monitored the motion of the heated patches on the water surface. An acoustic sensor was installed on the shutter to synchronise the shutter motion, wave probes and camera image capture. The absolute time reference between the thermal imagery and the wave probes was ±17 ms, determined by the camera framing rate.

3.4. Experiments

To validate the technique, the present experimental study was conducted to measure the water surface particle velocities and wave crest speeds for deep-water non-breaking and breaking waves with and without wind forcing. The TIV techniques were applied for monochromatic, C2N5 and C3N7 where C is the wave class and N is the number of waves in each temporal wave packet. A Class 3 wave packet (N=7) was selected to analyse under wind forcing. Before starting each experiment, the water surface was cleaned by generating waves to transport all slick material along the surface of the tank water to the beach. Once the water surface had been cleaned, the wave probes were immersed in the tank for approximately 1 hour to ensure their signal stability. To determine the conditions of maximum recurrence and marginal breaking for each group, paddle amplitudes were incrementally increased. The fetches of the group maximum were found carefully and the TIV system was sited over each location for subsequent measurement of the local wave characteristics and surface current. Initially, hot water droplets were used to generate hot spots on the water surface. As the droplets lost their thermal signatures quickly, it was necessary to deposit each droplet just before arrival of the wave crest.

Subsequently, the CO_2 laser system was synchronised to apply the heated patches to the water surface. The maximum wind speed at which traceable and sufficiently-localised heat patches could be tracked on the water surface was approximately 2.0 m/s. For wind speeds between 1.5 m/s and 2.0

m/s, hot spots rapidly expanded on the surface but they were still recognisable. Heated patches on wave crests for wind speeds higher than 2.0 m/s were not easy to track as they disappeared very quickly. Therefore, the maximum wind speed used for the present study was approximately 2.0 m/s.

4. Results and discussion

The TIV system was applied to non-breaking monochromatic and group waves. The displacement of heated patches at the crest of monochromatic and Class 3 waves are shown in figure 2. The water surface velocity was determined by tracking the leading edge of the heated patch in successive video frames and the wave crest speed was measured using the array of wave probes. The average and standard deviation from five measurements of water surface velocity (U_s), wave crest speed (C), wave amplitude (a) and wave number (k) for monochromatic waves (T = 0.7 sec) has been presented in table 1. The wave parameters were measured using the two arrays of wave probes.

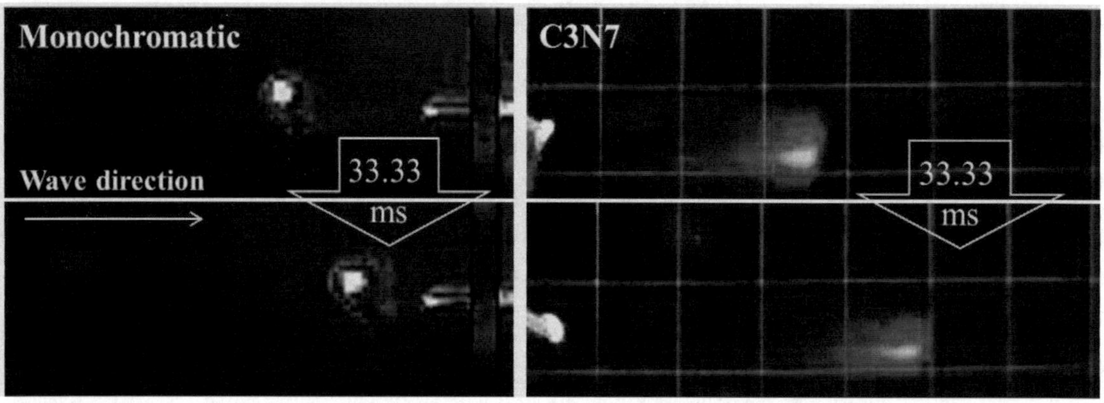

Figure 2. Tracking the hot spot on the wave crests of a monochromatic wave and a C3N7 wave packet for two successive frames using infrared imagery to measure the surface particle velocity.

The most challenging measurements required capturing crest surface velocities at the point of the group maximum. This required ensuring that the position of heat patches coincided with the crest concomitant with occurrence of the group maximum. Because the TIV method using the Thermos was not able to synchronise the droplets with the wave probes, it was not possible to have a precise measurement of the crest surface velocities for grouped waves. In addition, the experiment was conducted for breaking waves and it was found that the hot spots disappeared quickly and it was not possible to track those spots on the water surface. For wind-forced waves, the hot droplets were not trackable as they had lost their heat before reaching the water surface.

Table 1. The results of wave crest speed (C), wave amplitude (a), wave number (k), and water surface velocity (U_s) for a single case of monochromatic non-breaking waves (T = 0.7 s).

	C (m/s)	a (m)	k (m^{-1})	$a.k.C$	U_s (m/s)
Average	1.12	0.0092	7.84	0.081	0.085
Standard Deviation	0.014	0.00005	0.02	0.001	0.002

The TIV technique using the CO_2 laser improved the technique significantly for both breaking waves and wind-forced waves. Nonetheless, trial and error iteration was required to ensure that the position of heat patches coincided with a crest maximum event. The heated patches produced by the laser remained on the water surface longer than the droplets and were much more compact (approximately 4 mm in diameter). This allowed us to measure the water surface velocities before and after a crest maximum event as well as crest surface water velocity. The movement of a single heated patch on a C3N7 breaking wave is presented in figure 3. The location of the crest maximum is at the centre of the image. The first image shows the heated patch just before the breaking onset. Images (b) and (c) present the patch 33.33 ms and 66.66 ms thereafter. By digitising the front of the patches in successive frames and having the time between each frame, the water surface velocities could be calculated.

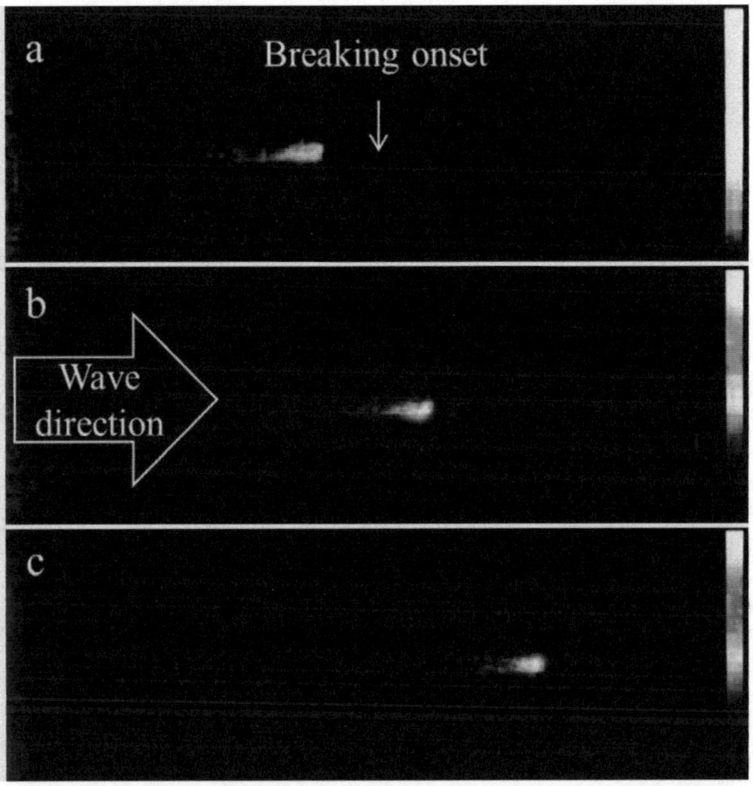

Figure 3. A heated patch generated by the CO_2 laser on a C3N7 wave packet (a) before the breaking onset, (b) and (c) on the breaking wave. The time between each frame is 33.33 ms.

After measuring the fluid velocities at the wave surface during the transit of a crest maximum event, these velocities could be interpolated in time to obtain a velocity at the crest that was immediately local in space to the crest maximum location. A polynomial curve was fitted to a sequence of such measurements (figure 4) to interpolate in space to obtain a single measurement of velocity at the crest group maximum. This entire sequence was repeated 7–10 times to quantify measurement repeatability. In each case, it was found that the water surface wave particle velocity increases to a maximum at the crest. The surface velocity for a marginal breaking C3N7 and C2N5 wave groups with the polynomial curve fitted to the velocities are illustrated in figure 4a. Distance in this figure indicates location relative to the crest maximum. As shown, the maximum water velocity occurs at a crest maximum event.

Further, the wave crest speeds were measured during transit of a crest maximum event. In contrast with surface velocity, the wave crest itself systematically slows down to a minimum speed at the crest maximum. The result of the crest speed measurement for the C3N7 wave group case shown in figure 3 and C2N5 wave groups with the polynomial curve fitted to the speeds are presented in figure 4b.

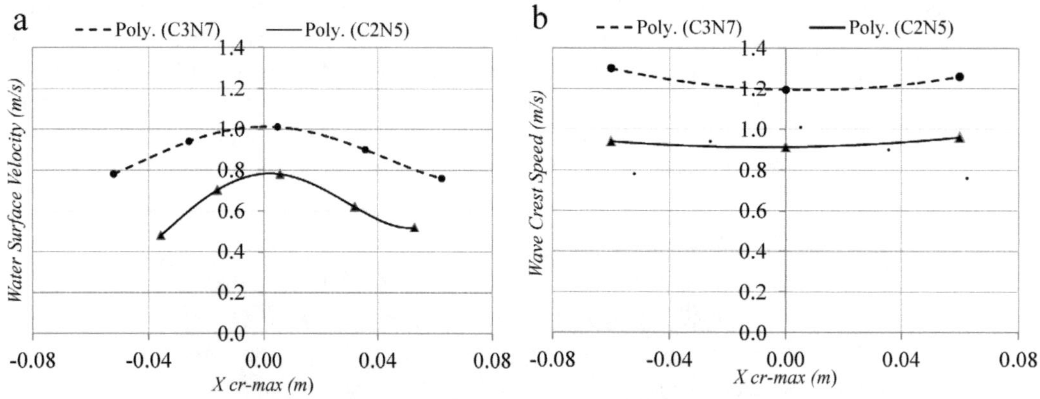

Figure 4. (a) Surface water particle velocities (m/s) and (b) wave crest speeds (m/s) in the vicinity of the crest maximum for marginal breaking C3N7 (dashed curve) and C2N5 (solid curve) wave groups. Horizontal axis is the distance relative to the location of crest maximum (m).

5. Summary and conclusions

In the present study a thermal image velocimetry technique using infrared imagery was developed to measure the water surface velocity. Heated droplets and a CO_2 laser were used to generate heated patches on the water surface. Unidirectional deep-water waves were generated in a 2D wind wave flume. Three wave types, i.e. monochromatic waves, bimodal spectrum (Class 2) and chirped wave packet (Class 3), were used for evaluation of wave breaking criteria.

It was shown that the TIV technique using heated water droplets can quantify the surface velocity for monochromatic, non-breaking waves without wind forcing.

A CO_2 laser system was developed to irradiate the water surface so as to improve the localisation of hot spot formation on the water surface. With this enhancement, the TIV technique can capture crest water velocities for unforced and wind-forced grouped waves.

These results show that the surface fluid velocity increases to a maximum at the crest maxima. At the same time, the wave profile crest speeds take on their minimum values at the crest maxima. The present results show that at the crest maximum location, the ratio of the water velocity U_s at the crest maximum to the wave profile speed C for marginal breaking waves is consistently less than unity.

References

[1] Peirson W L 1997 *Exp. Fluids* **23** 427
[2] Gray C and Greated C A 1998 *Opt. Lasers Eng.* **9** 265
[3] Siddiqui M H K *et al* 2001 *Phys. Fluids* **13** 1891
[4] Jensen A *et al* 2001 *Exp. Fluids* **30** 500
[5] Jessup A T and Phadnis K R 2005 *Meas. Sci. Technol.* **16** 1961
[6] Muste M *et al* 2005 *Flow Meas. Instru.* **16** 47
[7] Garbe C S *et al* 2007 *Transport at the Air–Sea Interface* eds Garbe C S *et al* (Heidelberg: Springer) 223-39
[8] Jessup A T *et al* 1997 *Nature* **385** 52
[9] Banner M L and Peirson W L 1998 *J. Fluid Mech.* **364** 115
[10] Zappa C J *et al* 2001 *J. Geophys. Res.* **106** 9385
[11] Stansell P and Macfarlane C 2002 *J. Phys. Oceanogr.* **32** 1269
[12] Peirson W L and Banner M L 2003 *J. Fluid Mech.* **479** 1
[13] Peirson W L *et al* 2014 *J. Fluid Mech.* **743** 399
[14] Banner M L *et al* 2014 *Phys. Rev. Lett.*, **112** 114502
[15] Barthelemy X *et al* 2015 *J. Fluid Mech.* under review
[16] Banner M L and Peirson W L 2007 *J. Fluid Mech.* **585** 93

IOP Conf. Series: Earth and Environmental Science **35** (2016) 012022 doi:10.1088/1755-1315/35/1/012022

Evaporation mitigation by floating modular devices

M M Hassan[1] **and W L Peirson**[1, 2]
[1] Water Research Laboratory, School of Civil and Environmental Engineering, University of New South Wales, 110 King Street, Manly Vale, NSW -2093, Australia
E-mail: m.hassan@wrl.unsw.edu.au
[2] New College, University of New South Wales, Kensington, NSW-2052, Australia
E-mail: bill.peirson@newcollege.unsw.edu.au

Abstract. Prolonged periods of drought and consequent evaporation from open water bodies in arid parts of Australia continue to be a threat to water availability for agricultural production. Over many parts of Australia, the annual average evaporation exceeds the annual precipitation by more than 5 times. Given its significance, it is surprising that no evaporation mitigation technique has gained widespread adoption to date. High capital and maintenance costs of manufactured products are a significant barrier to implementation. The use of directly recycled clean plastic containers as floating modular devices to mitigate evaporation has been investigated for the first time. A six-month trial at an arid zone site in Australia of this potential cost effective solution has been undertaken. The experiment was performed using clean conventional drinking water bottles as floating modules on the open water surface of 240-L tanks with three varying degrees of covering (nil, 34% and 68%). A systematic reduction in evaporation is demonstrated during the whole study period that is approximately linearly proportional to the covered surface. These results provide a potential foundation for robust evaporation mitigation with the prospect of implementing a cost-optimal design.

1. Introduction

Climate change is now recognised as one of the greatest potential challenges to human society and industry [1, p 12]. Water security is under threat in the arid and semi-arid regions during drought due to changes in precipitation or evaporation from open water bodies. For example, it is recognised that the eastern half of Australia was systematically drier from 1895–1948 than during 1948–2000. However, rainfall has been below long term averages across much of southeast Australia since 1997 [2].

In many arid and semi-arid regions, the primary source of water is from many small farm dams. Consequently, minimising water losses from such dams is fundamental to the ongoing economic viability of farm production in arid areas. Construction costs and the energy intensity of pumping water long distances may make supply of water from more well-watered regions infeasible. As Weeks [3] stated, "Considering the importance of evaporation in the water balance of reservoirs, it is surprising that so few detailed research projects have studied the problem."

The principal strategy in evaporation reduction is the minimisation of the water surface area that is in contact with the air as well as sheltering the water surface from the solar radiation [4, p 235]. Effective economic ways to reduce evaporation from large water bodies are yet to be developed [4, p 56].

Proposed open water evaporation mitigation techniques include wind sheltering by trees [5], (although, to our knowledge effectiveness has never been quantified); reservoir deepening [4, p 57],

Content from this work may be used under the terms of the Creative Commons Attribution 3.0 licence. Any further distribution of this work must maintain attribution to the author(s) and the title of the work, journal citation and DOI.
Published under licence by IOP Publishing Ltd

unquantified); sand storage dams / managed aquifer recharge; chemical monolayers; continuous coverings of the entire reservoir and floating modular devices.

Use of sand storage dams for conserving water has been one of the successful methods in mitigating evaporation [6–9]. Thesis studies show that evaporation can be eliminated if the water level is kept below 1 m [8] from the top of sand surface. However, the main limitations of such storages are their potentially low specific yield from the artificial aquifer [9]. Although attempts to increase the specific yield by improving gradation have been made by [9], specific yield never exceeds 47%. In addition, there may be risks (physical human and animal safety, contamination, poor water quality) associated with groundwater dams, some of which still may not yet have been identified [9]. Furthermore, managed aquifer recharge, groundwater dams may be compromised by large plants directly extracting water from the alluvium. Sand storage dams are used in ephemeral African streams [10].

Extensive research has been undertaken on retardation of evaporation using monolayers. Monolayers are thin chemical films that produce a diffusion barrier on the water surface [11]. Conventional monolayers are principally long chain alcohols. Vines and La Mer [12] demonstrated a reduction of evaporation using cetyl alcohol during field studies over a three-year deployment at Stephen's Creek Reservoir, Broken Hill; Umberumberka Reservoir, Broken Hill, NSW; Lake Corella, Mary Kathleen, Queensland and several other places throughout Australia. They demonstrated an overall reduction of evaporation, but monolayer efficiency reduces with wind speed; approximately 40% when the wind speed is below 5 mph; 20% as the wind speed approaches 10 mph; and no reduction if it exceeds 15 mph. The application of the slurry was a very time consuming and expensive task. Crow [13] proposed open wind baffles arranged in the grid network to hold the monolayer during strong winds. Crow [13] showed that using low length/height ratios for closed wind barriers would lead to 32% reduction in evaporation if wind speed exceeds 5 mph. Wind-generated, microscale breaking waves develop when wind speeds exceed approximately 5 m s^{-1}. Peirson and Banner [14] demonstrate that strong subduction of the water surface occurs beneath the spilling regions of these small waves. In addition, wind waves compress and stretch the open water surface potentially tearing any surface films present [15]. Therefore, monolayers cannot remain effective in mitigating evaporation once microscale breaking waves form. Most commonly used monolayer materials are also quite readily broken down by bacteria [16] and hence reapplication is necessary at regular intervals (typically 1–3 days) to maintain a surface cover. Monolayer materials adhere preferentially to dusts, which may reduce their efficiency [17].

Continuous physical coverings (solid roofs, plastic covers, Evap-mat, aquaspan, super-span, NetPro shade cloths, E-VapCap, etc.) should reduce evaporation by 100% [18–24]. Although reservoirs of approximately 0.16 km^2 have been covered in Australia, such undertakings are a major engineering enterprise [25]. The specific challenge associated with such activities is maintaining reservoir cover integrity as water levels rise and fall. High installation cost, potential water quality effects, and wind induced forces may also be significant.

Reduction of exposed surface area by floating barriers has been proposed [26–29]. However, to our knowledge no systematic study has been published to ensure the effect of degrees of surface coverings (from partly covered to almost fully covered). Moreover, high installation and maintenance cost due to breakage of continuous blocks have also hindered the development. Proposals that reservoir could be partially covered due to either economic or physical constraints has been challenged by Watts [30] on the basis that there are no data to support that it could be a linear relationship between partial coverage and evaporation mitigation.

This present contribution critically examines the potential of floating modular devices to mitigate evaporation from open water bodies. Evaporation mitigation was quantified during field experiments undertaken at an Australian inland site over a six-month period spanning autumn to spring. These included testing the effects of partial covering and determining the corresponding pan coefficients.

2. Field facility and methodology

2.1 Test facility

The experiments have been set up at Fowlers Gap Arid Zone Research Station (only Arid Zone Research Station in Australia, Latitude: 31.37°S, longitude: 141.659°E; altitude 181 m above MSL and 110 km north of Broken Hill Town), New South Wales, Australia. The topography is such that the station is surrounded by hills and the nearest coastline lies approximately 500 km southwest. Meteorological data were obtained from Fowlers Gap Weather Station located 60 m from the site. Polyethylene tanks of 240-L capacity (600 mm x 500 mm x 1000 mm) were fitted with a 150-mm diameter PVC well to maintain water levels in the tank. Three tanks were buried 100 mm apart and levelled with their lips approximately 150 mm higher than the adjacent ground level to minimise splashing from the tank [31]. The tanks were filled up with water to match the level of the adjacent ground surface and the free surfaces were covered with different numbers of floating modular devices. One tank was left uncovered as reference. A second was covered with eight 600 ml standard drinking water bottles to cover approximately 34% of the free surface area. The third tank was covered with sixteen bottles to cover approximately 68% surface area. The top of the tanks were covered with 12 mm x 12 mm open wired mesh to prevent access by birds or animals.

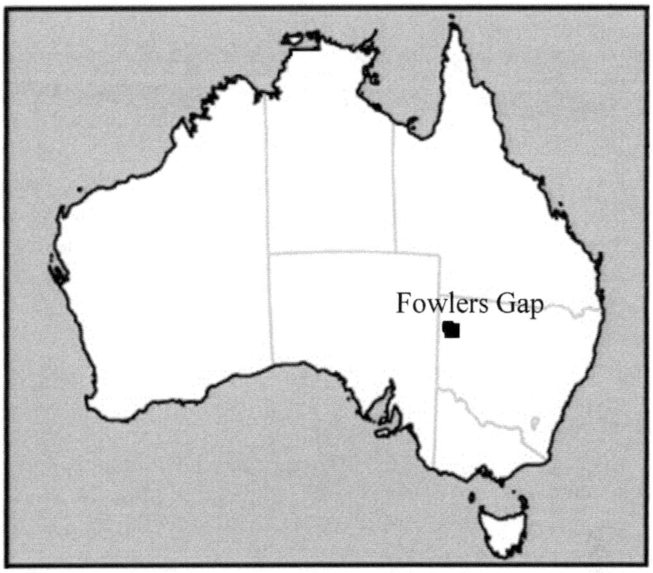

Figure 1. Map showing location of the experimental site

Figure 2. Picture showing experimental set-up

Acoustic sensors (mic+30/IU/TC/E, Microsonic GmbH & Co, Germany; measuring range 60–350 mm) were used to record the wells. They were mounted in 147mm perplex disc which was screwed within the tube and minimised humidity exchange between the well air chamber and the ambient air. The perplex discs provided just sufficient ventilation to avoid any pressure anomaly within the well cavity. Those sensors were connected to a National Instrument Data Acquisition system (PCI-6221) fitted to a standard personal computer. Rainfall was monitored by a laser precipitation monitor (LPM Model 5.4110.00.000 manufactured by Adolf Thiess GmbH & Co, Germany) connected to the same computer.

2.2 Test procedure

The water levels at different tanks were recorded in the form of voltages from LabVIEW signal Express at 0.5-s intervals on real time basis (one million measurements over 6 days for each sensor). HyperTerminal was used to record the rainfall amount and rain droplet size distribution in every minute. Regular monitoring using TeamViewer was undertaken remotely to maximise data capture. Tanks were refilled to ensure that the water levels remained in proximity of the adjacent ground surface level. Calibration equations were developed for each sensor to correlate between the voltages and distances at the start of the experiment and another calibration was performed at the end of the study period. Over the monitoring period, the maximum change in gain observed was less than 1.5%.

2.3 Data analysis

Diurnal effects were eliminated by determining daily evaporation rate by conventional linear regression [32] to the 24 hours of data from midnight to midnight. For each data point, 90% confidence intervals were also calculated.

Evaporation mitigation efficiency has been quantified as the evaporation rate from each covered tank (34% and 68% coverage) as a proportion of the corresponding reference (uncovered) tank.

3. Results and discussion

Figure 3 shows that evaporation rates from tanks and amount of rainfall received by the arid zone during the test period. Seasonal effects on evaporation rates from the reference tank are also evident.

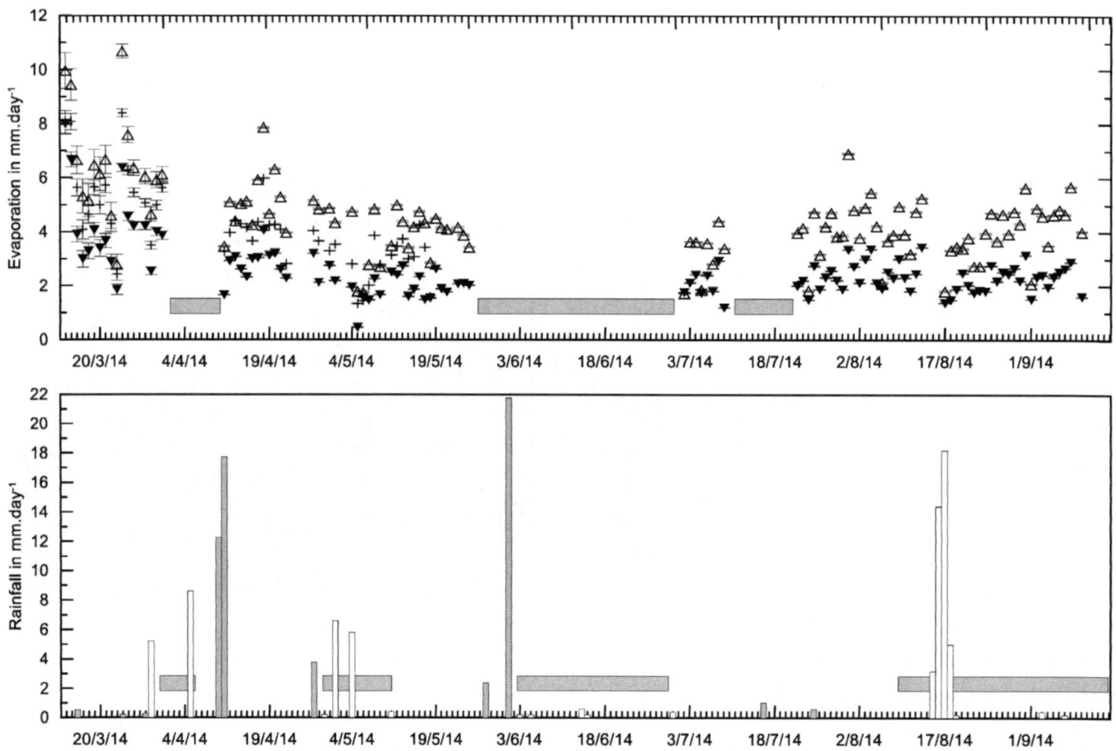

Figure 3. Top panel: Daily measurement of evaporation from the three tanks during the study period (March–September 2014). 90% confidence intervals are shown as error bars [upward pointing hollow triangles: Reference tank evaporation; downward pointing solid triangles: 68% covered tank; pluses: 34% covered tank]. Solid bars show periods of missing data. Due to sensor malfunction in the 34% covered tank, data after July have been discarded from all the analysis later than this period. Bottom panel: daily rainfall measurement from the LPM and the Fowlers Gap weather station.

During early autumn, evaporation rates were higher than rest of the period. During most of the study period, average daily evaporation was approximately 5 mm/hr. Due to an unknown problem in the sensor cable of the 34% covered tank after June 2014, evaporation measurements for that tank were excluded from rest of the analysis (figure 4).

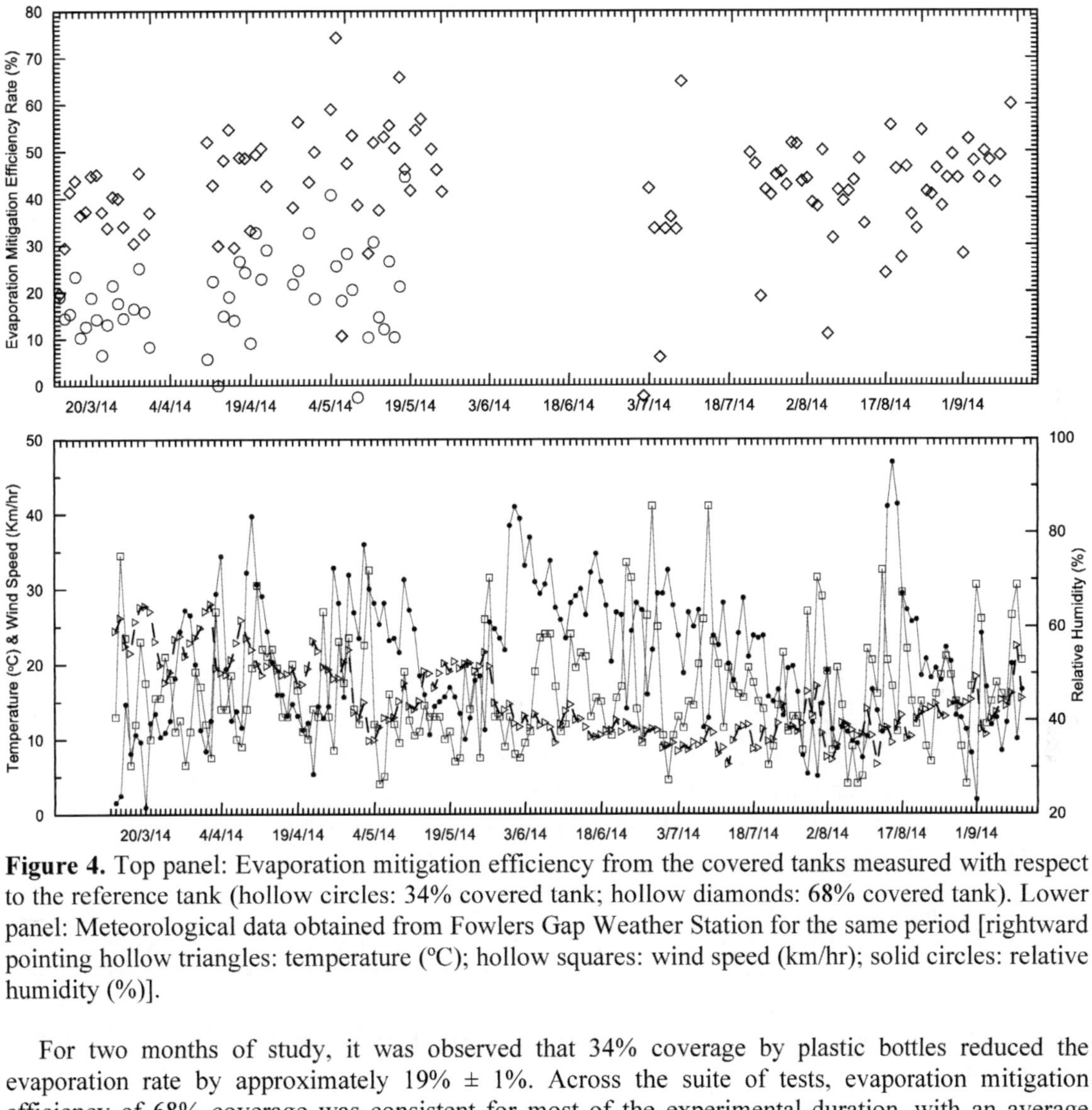

Figure 4. Top panel: Evaporation mitigation efficiency from the covered tanks measured with respect to the reference tank (hollow circles: 34% covered tank; hollow diamonds: 68% covered tank). Lower panel: Meteorological data obtained from Fowlers Gap Weather Station for the same period [rightward pointing hollow triangles: temperature (°C); hollow squares: wind speed (km/hr); solid circles: relative humidity (%)].

For two months of study, it was observed that 34% coverage by plastic bottles reduced the evaporation rate by approximately 19% ± 1%. Across the suite of tests, evaporation mitigation efficiency of 68% coverage was consistent for most of the experimental duration, with an average reduction rate of 43% ± 1% (figure 4).

By linear extrapolation of the evaporation reduction rate to 95% coverage, it is estimated that evaporation could be mitigated by 58% for a single layer of 65-mm drinking water bottles. The mean measured mitigation rates at this site for the partially covered tanks are as shown in figure 5 in comparison with the recent study by Hassan *et al.* [33].

IOP Conf. Series: Earth and Environmental Science **35** (2016) 012022 doi:10.1088/1755-1315/35/1/012022

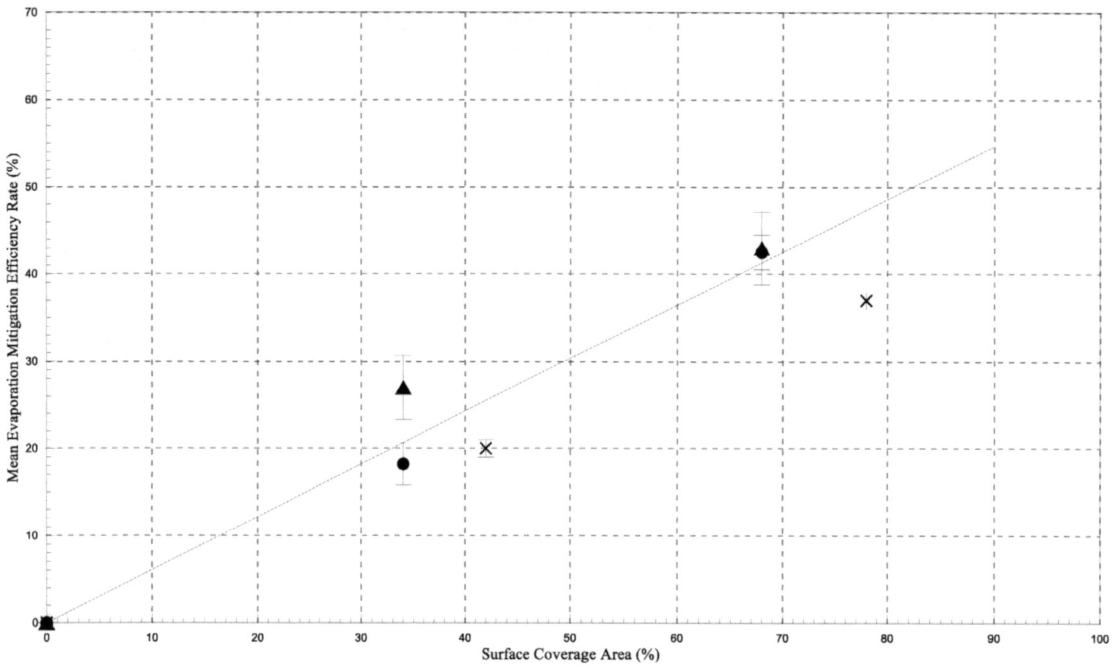

Figure 5. Mean evaporation mitigation efficiency rate as a function of water surface coverage area. 90% confidence limits are shown as error bars. Solid circles: Fowlers Gap; solid triangles: WRL, Sydney [33]; crosses: Ki Downs, Mildura [33]

4. Conclusions and recommendations

The effectiveness of near zero cost and widely available recycled material in the form of floating modules to reduce evaporation has been performed in the field. The results obtained are consistent with the conclusions of Hassan *et al.* [33] that evaporation loss reduces approximately linearly in proportion with the degree surface covering. Long-term deployments have demonstrated that the results are robust across a range of climatic conditions.

Full scale field testing on real farm dams and possible water quality impacts in partially covered tanks should be investigated.

Acknowledgments

The authors gratefully acknowledge the assistance of Dr. Keith Leggett, Director, Fowlers Gap Research Station, who permitted us to install the experiment at Fowlers Gap Research station and also helped maintain the experiment throughout the study period. The authors also express appreciation for the construction assistance by Bryce Neyland, Nicholas Fiddis, Larry Paice and Robert Jenkins of the Water Research Laboratory.

References

[1] Pachauri R and Reisinger A 2007 *IPCC Fourth Assessment Report* (Geneva: IPCC)
[2] Murphy B F and Timbal B 2008 *Int. J. Climatol.* **28** 859
[3] Weeks W D 1983 *The Calculation of Evaporation in Queensland* (Institution of Engineers Australia, Queensland Division) QBG 1756
[4] Pereira L S *et al* 2002 *Coping With Water Scarcity* (Paris: UNESCO International Hydrological Programme)
[5] Hipsey M 2002 Using windbreaks to reduce evaporation from farm dams *Farmnote* 72/2002
[6] Hellwig D 1973 *J. Hydrology* **18** 317

[7] Wipplinger O 1953 *The Storage of Water in Sand: An Investigation of the Properties of Naturaland Artificial Sand Reservoirs and of Methods of Developing Such Reservoirs* (PhD Thesis, Stellenbosch University)
[8] Wipplinger O 1958 *Storage of Water in Sand, South West Africa [Namibia]* Administration Water Affairs Branch, Windhoek, Namibia
[9] Peirson W L *et al* 2010 Evaporation mitigation by storage in rock and sand *Gas Transfers at Water Surfaces* (Kyoto: Kyoto University Press)
[10] Nilsson A 1988 *Groundwater Dams for Small-Scale Water Supply* (London: Intermediate Technology Publications Ltd)
[11] Mcjannet D *et al* 2008 Evaporation reduction by monolayers: overview, modelling and effectiveness *Technical Report, UWSRA, ISSN 1836-5566*
[12] Vines R and La Mer V 1962 *Retardation of Evaporation by Monolayers* (Academic Press)
[13] Crow F R 1964 *The Effect of Wind on Evaporation Suppressing Films and Methods of Modifications* (Oklahoma State University)
[14] Peirson W L and Banner M L 2003 *J. Fluid Mech.* **479** 1-38
[15] Peirson W L *et al* 2014 *J. Fluid Mech.* **743** 399
[16] Chang S *et al* 1962 Effect of bacterial decomposition of hexadecanol and octadecanol in monolayer films on the suppression of evaporation loss of water *Retardation of Evaporation by Monolayers: Transport Processes* (New York: American Chemical Association)
[17] Speed V J 1967 *An Engineering Evaluation on Evaporation Suppression by Mono-Layers* (PhD thesis, Civil Engineering)
[18] Craig I *et al* 2005 Controlling evaporation loss from water storages *NCEA report, University of Southern Queensland, Department. of Agriculture, Civil and Environmental Engineering* doi: 1000580/1
[19] Craig I P 2005 Loss of storage water due to evaporation - a literature review *NCEA report, University of Southern Queensland, Department of Agriculture, Civil and Environmental Engineering* doi: 1000580/0
[20] Alvarez V M *et al* 2006 *J. Agr. Water Manag.* **84** 229
[21] Considine M L 2007 Putting a thin lid on evaporation losses *ECOS, CSIRO Publishing*
[22] Finn N and Barnes S 2007 The benefits of shade-cloth covers for potable water storages *CSIRO Textile & Fibre Technology, CSIRO Gale Pacific* Victoria, Australia
[23] Gallego-Elvira B *et al* 2009 Reducing Evaporation Losses from Agricultural Reservoirs by Means of Shade-cloth Covers under a Semiarid Climate *Agricultural Technologies In a Changing Climate: The 2009 CIGR International Symposium of the Australian Society for Engineering in Agriculture. Brisbane, Queensland: Engineers Australia* 147-155
[24] Martínez Á V *et al* 2010 *Spanish J. of Agr. Res.* **8** doi: 10.5424/sjar/201008S2-1355
[25] Cox W *et al* 2000 Innovative Reservoir Design andConstruction at Potts Hill *Ensuring Save, Secure Water Supply, Water TECH, enviro., Sydney, 9-13 April*
[26] Cooley K R 1983 *J. Irrig. Drain. Eng.* **109** 89
[27] Burston I A and Akbarzadeh A 1995 Fresh water storage–evaporation research, reducing loss from evaporation *Report to the Cotton Research and Development Corporation, Sainty.*
[28] Burston I 2002 *Conservation of Water from Open Storages by Minimising Evaporation* (PhD thesis, University of Queensland)
[29] Segal L and Burstein L 2010 *Water Resources Management* **24** 129
[30] Watts P 2005 Scoping Study: Reduction of Evaporation from Farm Dams *Final Report to the National Program for Sustainable Irrigation, Feedlot Services Australia Pty Ltd*Toowoomba.
[31] Yarin A L 2006 *Annu.Rev. Fluid Mech.* **38** 159
[32] Walpole E R 1974 *Introduction to Statistics* (New York: Macmillan)
[33] Hassan M M *et al* 2015 *J. Hydrol.* **530** 742